The Role of Microbes in Autoimmune Diseases

Ahed J. Alkhatib

The Role of Microbes in Autoimmune Diseases

New Mechanisms of Microbial Initiation of Autoimmunity

Springer

Ahed J. Alkhatib
Department of Legal Medicine, Toxicology and Forensic Medicine
Jordan University of Science and Technology
Irbid, Jordan

ISBN 978-981-19-1161-3 ISBN 978-981-19-1162-0 (eBook)
https://doi.org/10.1007/978-981-19-1162-0

This Springer imprint is published by the registered company Springer Nature Singapore Pte Ltd.
The registered company address is: 152 Beach Road, #21-01/04 Gateway East, Singapore 189721, Singapore

I would like to dedicate this book to the great people who have had impact on my academic path; Professor Oleg Yurevich Latyshev, the President of International Mariinskaya Academy (IMA), for his great efforts in uniting academicians from the world in IMA.
I would also like to dedicate this book to Professor Tuweh Negus Prince Gadama the Great for his real great achievement in academia, education, and politics.
I would also like to dedicate this book to all my family members: wife and sons. Also, I would like to dedicate this book to the soul of my parents.

Preface

Writing literature is considered easier than writing in science. In literature, the writer may easily claim an opinion, idea, or even having a perception. It is not a matter to agree or disagree with the writer. At school and university ages, it was a habit to write some poems and philosophical ideas. My dreams were far away from writing books, because a good level of maturity is required to transform ideas into scientific texts. During that time, I worked to acquire scientific mentality that gives ideas and accepts criticism. I was planning to start the stage of book writing after being retired. It was my ambition to see my books translated into different languages. At the end of 2017, I received an e-mail from a publishing house to transform one of my articles into a book. The first book was about the molecular and physiological roles of estrogen receptor. This was followed by another book *White Matter and Disease: Does Brain have a Role in Initiating Disease*. I was glad that both books were released in nine different languages.

The idea of this book *The Role of Microbes in Autoimmune Diseases* came into my mind as a response to a question how fungi are interacting with the presence of estrogen? Various theories tried to suggest answers including the presence of estrogen binding protein expressed in the yeast cell itself. The idea was further developed to become an explanation of the development of autoimmunity by microbes.

I am glad that the idea of this book has been accepted by Springer nature and to make this book.

This book is mainly important for readers as it introduces new perceptions for the initiating process of autoimmune diseases. I have started thinking of microbes from different points of view for a long time. Beyond the structural variations between microbial cells and host cells, I reached to the conclusion that microbes act functionally as host cells, including human cells. There are similar mechanisms in cell divisions including p53, estrogen receptor, BCL2, and others. Similar neurotransmitters are released by microbes and host cells.

Proteins included in cell cycle division are thought to be conserved and not exposed to immune cells. Being produced by microbes implies to be exposed to immune cells in different areas of the body and development of autoantibodies and as a result autoimmune disease. It seems that this is a new approach explaining the development of autoimmune diseases and adds a new piece of information to the existing literature.

This book consists of seven chapters that cover immunology subjects in general and autoimmunity.

Irbid, Jordan Ahed J. Alkhatib

Acknowledgments

I would like to thank everyone who helped me in the production of this book. This book would not have been completed if the invitation and follow-up by Springer Nature editors Bhavik Sawhney and Immaculate Jayanthi had not initiated and continued.

I would also like to thank my wife for her continuous support, particularly in the COVID-19 period. I would like to thank my daughter, Dr. Ilham Alkhatib, who helped me in discussing different parts of this book and for suggestions in making the figures in this book.

Contents

Abbreviations

AA	Alopecia areata
ACPAs	Anticitrullinated protein antibodies
AD	Atopic dermatitis
ADCC	Antibody-dependent cytotoxicity
AICy	Autoimmune cytopenias
AID	Autoimmune disorders
AIDP	Acute inflammatory demyelinating polyradiculoneuropathy
AIG	Autoimmune gastritis
AIH	Autoimmune hepatitis
AITD	Autoimmune thyroid diseases
Anti-CarP	Anti-carbamylated protein antibodies
APS	Antiphospholipid syndrome
C. albicans	*Candida albicans*
CIA	Collagen-induced arthritis
CIDP	Chronic inflammatory demyelinating polyneuropathy
CMML	Chronic myelomonocytic leukemia
CNS	Central nervous system
DCs	Dendritic cells
EPHB2	Ephrin type B receptor 2
ER	Estrogen receptor
GWAS	Genome-wide association studies
HSCT	Hematopoietic stem cell transplants
IgA	Immunoglobulin A
LGI1	Anti-leucine-rich glioma inactivated
MAS	Multiple autoimmune disorders
MBL	Mannose-binding lectin
MDA5	Melanoma differentiation associated gene 5
MDS	Myelodysplastic syndromes
MGUS	Monoclonal gammopathy of unknown significance
MMN	Multifocal motor neuropathy
MOG	Myelin oligodendrocyte glycoprotein
MS	Multiple sclerosis
NBOD2	Nucleotide-binding oligomerization domain-containing protein 2

NLRs	NOD-like receptors
NMOSD	Neuromyelitis optica spectrum disease
PNS	Peripheral nervous system
Ps	Psoriasis
RA	Rheumatoid arthritis
RF	Rheumatoid factor
ROS	Reactive oxygen species
SDB	Sabouraud Dextrose Broth
SLE	Systemic lupus erythematosus
SS	Sjoègren's syndrome
TLR5	Toll-like receptor 5
TNF	Tumor necrosis factor
Treg	Regulatory T cells

List of Figures

Immunology: Principles and Applications 1

1.1 Immunology: Principles and Applications

1.1.1 Introduction to Immunology

The purpose of this chapter is to provide the reader an idea about the principles and applications of immunology. Immunology can be defined from various perspectives such as medicine, biology, and language. The definitions have common features as a branch of biological sciences that concerns approaches followed by the body to defend itself from foreign substances such as infectious agents. The main feature of the immune system is its ability to distinguish self and non-self-antigens (Napier 2012; Shahsavar et al. 2017; Yasin et al. 2019).

The immune system is consisted of structures that are specialized to offer protection against a variety of interfering aspects including microbes such as bacteria, viruses, parasites, allergens, toxins, malignant cells, and others (González et al. 2019).

The function of the immune system can be viewed from two perspectives according to specificity as either innate immunity (non-specific) or adaptive immunity (specific) (Kroker 1999; Jamieson 2010; Nicholson 2016). Innate immune system acts through its fundamental cells, neutrophils, and macrophages to remove infections, because these cells are the first to be in contact with pathogens at scene of contact such as skin lesions (de Melo Cruvinel et al. 2010; Parihar et al. 2010). Neutrophils and macrophages are both considered efficient killers of pathogens through the process of secretion of certain chemicals that induce destructive effects of pathogens. Illustrating examples are digesting enzymes. Following the killing process, macrophages and neutrophils eat or swallow up the killed pathogens and their related constituents, a process known as phagocytosis (Pollock et al. 1995; Hart et al. 2005; Silva 2010). If these processes are not adequate to destroy pathogens, the role of lymphocytes follows. The second line of defense mechanism, adaptive immunity, plays a major role through memory function of lymphocytes which

A. J. Alkhatib, *The Role of Microbes in Autoimmune Diseases*,
https://doi.org/10.1007/978-981-19-1162-0_1

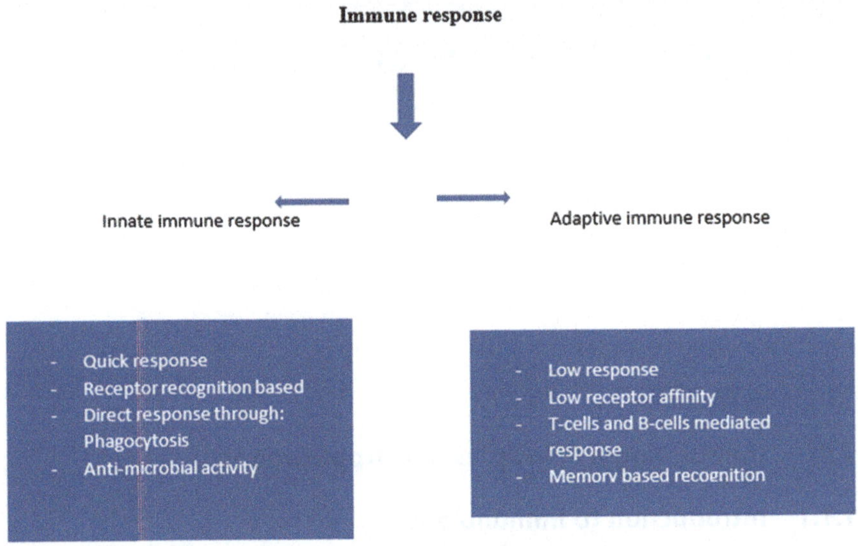

Fig. 1.1 Differences in immune response (innate and adaptive)

permits the immune system to create specific reactions and to memorize different types of infection which enable the immune system to treat reinfections with more efficient mechanisms (Heine and Lien 2003; Rossi and Young 2005; Hoffmann and Akira 2013) (Fig. 1.1).

1.2 Cellular Components of Immune System

All blood cells, including red blood cells, platelets, and white blood cells, are derived from the same precursor cells or progenitor cells in the bone marrow. These precursor cells are also called the hematopoietic stem cells. When these stem cells can generate various types of blood cells, they are called pluripotent hematopoietic stem cells (Nathan et al. 1978; Bao et al. 2019; Jacobsen and Nerlov 2019).

Granulocytes, macrophages, and dendritic cells, in addition to mast cells, are derived from myeloid progenitor. Macrophages are considered part of phagocytes due to immune system existing in different tissues and crucially act as important players of innate immunity (Weissman 2000; Miranda and Johnson 2007; Christiano et al. 2019). Macrophages are considered mature form of monocytes that differentiate to macrophages within tissues (Hoeffel and Ginhoux 2018; Olingy et al. 2019; Wolf et al. 2019). Other immune cells, dendritic cells act to capture antigens and then expose it for lymphocytes to be recognized. The journey of immature dendritic cells starts from blood and ends in the tissues in which they carry out their functions as phagocytosis and macropinocytic. Following being in contact with pathogens, they become mature and are more likely to migrate to lymph nodes (Fogg et al. 2006; Auffray et al. 2009; Yanez et al. 2017) (Fig. 1.2).

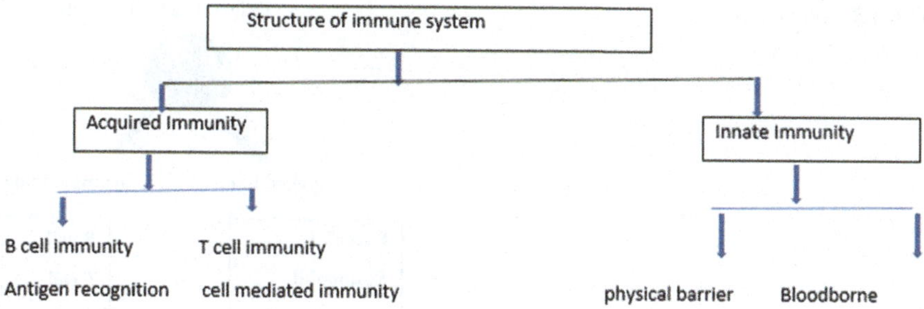

Fig. 1.2 Cellular components of immune system

Mast cells have no well-known precursors and usually complete their differentiation in tissues. They reside close to blood vessels, particularly small ones, and upon activation, secrete chemicals that impact vascular permeability. Their main roles are exerted to mediate allergic reactions, and they act to partly contribute to protect the surfaces of mucosa from pathogens (Steiner et al. 1985; Strauss-Albee et al. 2014; Krystel-Whittemore et al. 2016).

The granulocytes present important immune cells and because of having dense granules in their cytoplasm, they acquired their name as granulocyte. Three types of granulocytes have been identified: neutrophils, eosinophils, and basophils (Loktionov 2019). Neutrophils are considered as phagocytic cells that are the most abundant and important cells in innate immunity (Selders et al. 2017). Deficiency in neutrophil either in number or function is associated with infection burden (Kumar and Sharma 2010). Eosinophils are important immune cells defending against parasitic infections (Manali et al. 2018). Basophils have common features with eosinophils and mast cells, although distinguished features have been reported (Stone et al. 2010).

Lymphocytes are derived from lymphoid progenitor and are of two main types: B lymphocytes or B cells, and T lymphocytes or T cells (Khosrotehrani et al. 2008). B cells are differentiated upon activation into plasma cells which release antibodies. T cells comprise two main cellular populations: cytotoxic T cells that attack and kill virus-infected cells. On the other hand, the second class of T cells can be differentiated into other cells that activate cells including B cells and macrophages (Xiao et al. 2016; Sanz et al. 2019).

Lymphocytes are characterized by their ability to exhibit a specific immune response against foreign bodies (antigens) due to process of maturity that involves the existence of unique antigen receptor with great diversity in antigen binding sites (Tipton et al. 2015; Jorge et al. 2019).

The B-cell antigen receptor (BCR) presents the antibody that will be released by B cells after being activated and differentiated into plasma cells (Nutt et al. 2015). Collectively speaking, antibodies are molecules that are known as immunoglobulins (Ig). On the other hand, T-cell antigen receptor (TCR) closes to immunoglobulin, and it is directed to bind antigens derived from foreign sources (Richard et al. 2019).

Fig. 1.3 Immune cells

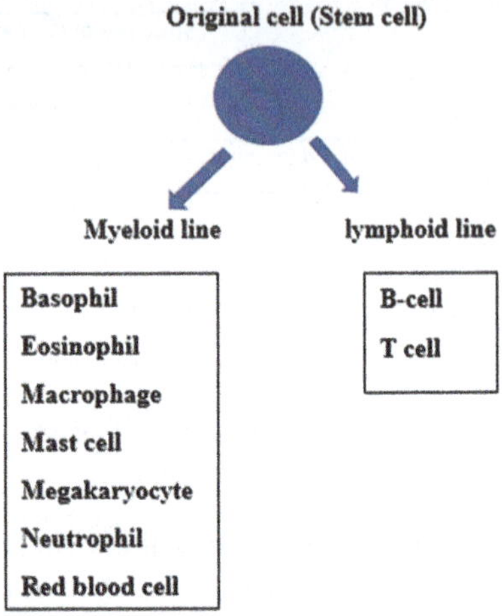

Another subpopulation of lymphoid cells is known as natural killer cells. These cells act independently of antigen specific receptors and consequently are considered part of innate immune system. These cells are large lymphocytes that have cytotoxic granules and can attack and kill invalid cells including tumor cells as well as virus-infected cells (Sarhan et al. 2016; Ziqing et al. 2019) (Fig. 1.3).

1.3 Principles of Immunological Reactions

It is the job of white blood cells to interact with infections. The first episodes of immunological interactions are not specific and involve the innate immune responses through the actions of neutrophils and macrophages, an illustrative example is an insect bite (Nicholson 2016). If innate immune responses were not adequate, another more advanced responses will be triggered during lymphocytes. Here, two characteristics are involved: adaptation and memory. Immune cells express large number of receptors that help to interact with their environment; a matter that guides how genes are utilized in immunological response (Nicholson 2016). Immune system is continuously encountering challenges that require to be well manipulated (Fig. 1.4). Accordingly, the immune system must recognize these changes surrounding it and to develop appropriate responses (Serra et al. 2017). One of these challenges is the discrimination. The relationship between immune system with its environment is complicated. Contact with new bodies is not always harmful, but it can be dangerous. The efficacy of the immune system requires the recognition ability of differences such as self from non-self-antigens and harmless non-self from dangerous non-self (Nicholson 2016).

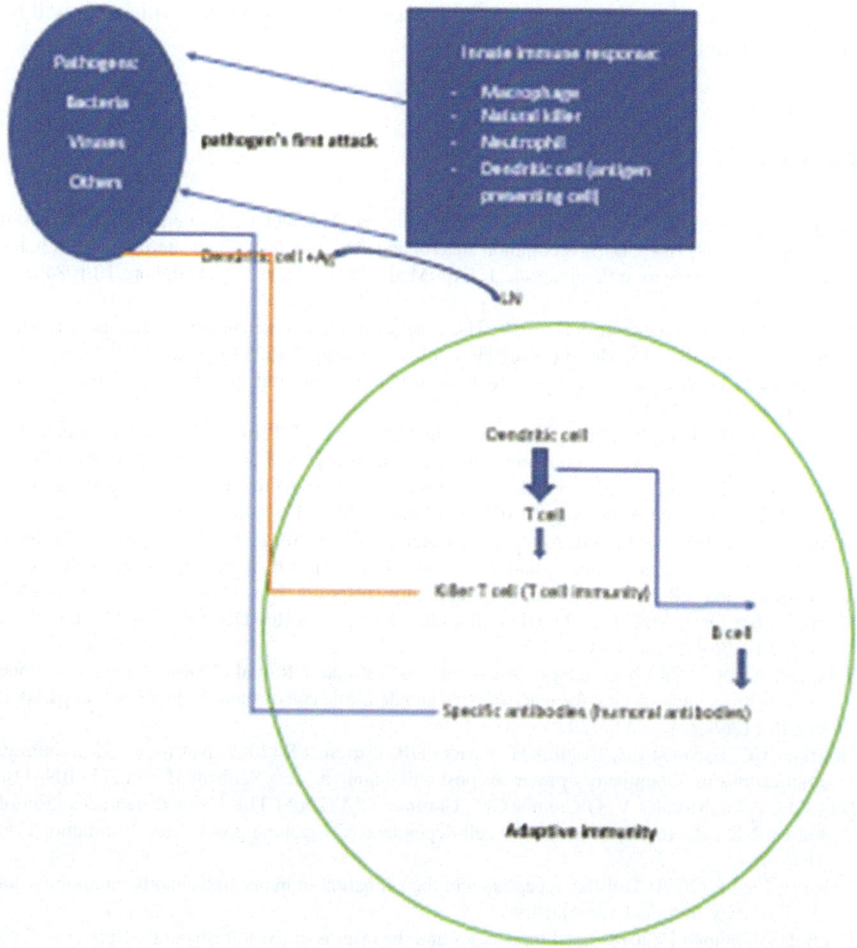

Fig. 1.4 An illustration of the function of immune system towards treating invading pathogens

1.4 Clinical Implications of Immunology

Immunology, from a personal perspective of view, may be the most important in biology and medical sciences. This claim is due to several factors among which its clinical implications in diagnosis including, but not restricted to, immunohistochemistry, western blot, flow cytometry, and any other technique involving the use of antibodies to identify materials (Chan 2000; Makki 2016). The second aspect explaining clinical implications of immunology is its involvement indirectly or directly in the development of various diseases. Indirect involvement of immunology in disease progression involves, but not limited to, diabetes, cancers, and neurological disorders such as multiple sclerosis (Filippi et al. 2018). Direct diseases related to immunology include several diseases such as hypersensitivities that will be discussed in detail in this book.

References

Auffray C, Fogg DK, Narni-Mancinelli E, Senechal B, Trouillet C, Saederup N et al (2009) CX3CR1+ CD115+ CD135+ common macrophage/DC precursors and the role of CX3CR1 in their response to inflammation. J Exp Med 206:595–606. https://doi.org/10.1084/jem.20081385

Bao EL, Cheng AN, Sankaran VG (2019) The genetics of human hematopoiesis and its disruption in disease. EMBO Mol Med 11(8):e10316. https://doi.org/10.15252/emmm.201910316

Chan JK (2000) Advances in immunohistochemistry: impact on surgical pathology practice. Semin Diagn Pathol 17:170–177

Christiano MVB, Fock RA, Hastreiter AA, Reutelingsperger C, Perretti M, Paredes-Gamero EJ, Farsky SHP (2019) Extracellular annexin-A1 promotes myeloid/ granulocytic differentiation of hematopoietic stem/ progenitor cells via the Ca2+/MAPK signalling transduction pathway. Cell Death Dis 5:135. https://doi.org/10.1038/s41420-019-0215-1

de Melo Cruvinel W, Júnior DM, Araújo JAP, Catelan TTT, de Souza AWS, da Silva NP, Andrade LEC (2010) Immune system—part I fundamentals of innate immunity with emphasis on molecular and cellular mechanisms of inflammatory response. Bras J Rheumatol 50(4):434–461

Filippi M, Bar-Or A, Piehl F et al (2018) Multiple sclerosis. Nat Rev Dis Primers 4:43. https://doi.org/10.1038/s41572-018-0041-4

Fogg DK, Sibon C, Miled C, Jung S, Aucouturier P, Littman DR et al (2006) A clonogenic bone marrow progenitor specific for macrophages and dendritic cells. Science 311:83–87. https://doi.org/10.1126/science.1117729

González CC, Escorcia LR, Castillo JT, Arrieta DB, Carrero LS (2019) Nutrition and immunity: implications in inflammatory processes post chikungunya. Acta Sci Nutr Health 3(5):109–118

Hart OM, Athie-Morales V, O'Connor GM, Gardiner CM (2005) TLR7/8-mediated activation of human NK cells results in accessory cell-dependent IFN-gamma production. J Immunol 175: 1636–1642

Heine H, Lien E (2003) Toll-like receptors and their function in innate and adaptive immunity. Int Arch Allergy Immunol 130:180–192

Hoeffel G, Ginhoux F (2018) Fetal monocytes and the origins of tissue-resident macrophages. Cell Immunol 330:5–15. https://doi.org/10.1016/j.cellimm.2018.01.001

Hoffmann J, Akira S (2013) Innateimmunity. Curr Opin Immunol 25:1–3

Jacobsen SEW, Nerlov C (2019) Haematopoiesis in the era of advanced single-cell technologies. Nat Cell Biol 21:2–8

Jamieson M (2010) Imagining 'reactivity': allergy within the history of immunology. Stud Hist Philos Biol Biomed Sci 30:273–296

Jorge A, Wallace ZS, Zhang Y, Lu N, Costenbader KH, Choi HK (2019) All-cause and cause-specific mortality trends of end-stage renal disease due to lupus nephritis from 1995 to 2014. Arthritis Rheumatol 71:403–410. https://doi.org/10.1002/art.40729

Khosrotehrani K, Leduc M, Bachy V, Huu SN, Oster M, Abbas A, Uzan S (2008) Pregnancy allows the transfer and differentiation of fetal lymphoid progenitors into functional T and B cells in mothers. J Immunol 180:889–897

Kroker K (1999) Immunity and its other: the anaphylactic selves of Charles Richet. Stud Hist Philos Biol Biomed Sci 30:273–296

Krystel-Whittemore M, Dileepan KN, Wood JG (2016) Mast cell: a multi-functional master cell. Front Immunol 6:620. https://doi.org/10.3389/fimmu.2015.00620

Kumar V, Sharma A (2010) Neutrophils: Cinderella of innate immune system. Int Immunopharmacol 10:1325–1334

Loktionov A (2019) Eosinophils in the gastrointestinal tract and their role in the pathogenesis of major colorectal disorders. World J Gastroenterol 25(27):3503–3526

Makki JS (2016) Diagnostic implication and clinical relevance of ancillary techniques in clinical pathology practice. Clin Med Insights Pathol 9:5–11. https://doi.org/10.4137/CPath.S32784

Manali M, Paige L, Shigeharu U (2018) Eosinophil extracellular traps and inflammatory pathologies—untangling the web! Front Immunol 9:2763

Miranda MB, Johnson DE (2007) Signal transduction pathways that contribute to myeloid differentiation. Leukemia 21:1363–1377

Napier AD (2012) Nonself help: how immunology might reframe the enlightenment cult. Anthropologie 27:122–137

Nathan DG, Clarke BJ, Hillman DG, Alter BP, Housman DE (1978) Erythroid precursors in congenital hypoplastic (diamond-Blackfan) anemia. J Clin Invest 61:489–498

Nicholson LB (2016) The immune system. Essays Biochem 60:275–301. https://doi.org/10.1042/EBC20160017

Nutt SL, Hodgkin PD, Tarlinton DM, Corcoran LM (2015) The generation of antibody-secreting plasma cells. Nat Rev Immunol 15:160–171. https://doi.org/10.1038/nri3795

Olingy CE, Dinh HQ, Hedrick CC (2019) Monocyte heterogeneity and functions in cancer. J Leukoc Biol 106(2):309–322. https://doi.org/10.1002/JLB.4RI0818-311R

Parihar A, Eubank TD, Doseff AI (2010) Monocytes and macrophages regulate immunity through dynamic networks of survival and cell death. J Innate Immun 2:204–215

Pollock JD, Williams DA, Gifford MA et al (1995) Mouse model of X-linked chronic granulomatous disease, an inherited defect in phagocyte superoxide production. Nat Genet 9:202e9

Richard DP, Baltimore D, Li G (2019) The cellular immunotherapy revolution: arming the immune system therapy for precision. Trends Immunol 40(4):292–309

Rossi M, Young JW (2005) Human dendritic cells: potent antigen-presenting cells at the crossroads of innate and adaptive immunity. J Immunol 175:1373–1381

Sanz I, Wei C, Jenks SA, Cashman KS, Tipton C et al (2019) Challenges and opportunities for consistent classification of human B cell and plasma cell populations. Front Immunol 10:2458. https://doi.org/10.3389/fimmu.2019.02458

Sarhan D, Cichocki F, Zhang B, Yingst A, Spellman SR, Cooley S, Verneris MR, Blazar BR, Miller JS (2016) Adaptive NK cells with low TIGIT expression are inherently resistant to myeloid-derived suppressor cells. Cancer Res 76:5696–5706

Selders GS, Fetz AE, Radic MZ, Bowlin GL (2017) An overview of the role of neutrophils in innate immunity, inflammation and host-biomaterial integration. Regener Biomater 4(1):55–68. https://doi.org/10.1093/rb/rbw041

Serra MB, Barroso WA, Neves N, da Silva S, Silva d N, Borges ACR, Abreu IC, da Rocha Borges MO (2017) From inflammation to current and alternative therapies involved in wound healing. Int J Inflamm 2017:3406215. https://doi.org/10.1155/2017/3406215

Shahsavar F, Varzi AM, Ahmadi SA (2017) A genomic study on distribution of human leukocyte antigen (HLA)-a and HLA-B alleles in Lak population of Iran. Genom Data 11:3–6. https://doi.org/10.1016/j.gdata.2016.11.012

Silva MT (2010) When two is better than one: macrophages and neutrophils work in concert in innate immunity as complementary and cooperative partners of a myeloid phagocyte system. J Leukoc Biol 87:93–106

Steiner DR, Gonzalez NC, Wood JG (1985) Mast cells mediate the microvascular inflammatory response to systemic hypoxia. J Appl Physiol 94(1):325–334. https://doi.org/10.1152/japplphysiol.00637.2002

Stone KD et al (2010) IgE, mast cells, basophils, and eosinophils. J Allergy Clin Immunol 125(2):73–80

Strauss-Albee DM, Horowitz A, Parham P, Blish CA (2014) Coordinated regulation of NK receptor expression in the maturing human immune system. J Immunol 193(10):4871–4879. https://doi.org/10.4049/jimmunol.1401821

Tipton CM, Fucile CF, Darce J, Chida A, Ichikawa T, Gregoretti I et al (2015) Diversity, cellular origin and autoreactivity of antibody-secreting cell population expansions in acute systemic lupus erythematosus. Nat Immunol 16:755–765. https://doi.org/10.1038/ni.3175

Weissman IL (2000) Translating stem and progenitor cell biology to the clinic: barriers and opportunities. Science 287:1442–1446

Wolf AA, Alberto Y, Barman PK, Goodridge Helen S (2019) The ontogeny of monocyte subsets. Front Immunol 10:1642. https://doi.org/10.3389/fimmu.2019.01642

Xiao X, Lao X-M, Chen M-M, Liu R-X, Wei Y, Ouyang F-Z et al (2016) PD-1hi identifies a novel regulatory B-cell population in human hepatoma that promotes disease progression. Cancer Discov 6:546–559. https://doi.org/10.1158/2159-8290.CD-15-1408

Yanez A, Coetzee SG, Olsson A, Muench DE, Berman BP, Hazelett DJ et al (2017) Granulocyte-monocyte progenitors and monocyte-dendritic cell progenitors independently produce functionally distinct monocytes. Immunity 47:890–902e894. https://doi.org/10.1016/j.immuni.2017.10.021

Yasin SA, Shahsavar AF, Rezaian KAJ (2019) An introduction to the role of immunology in medical anthropology andmolecular epidemiology. Biomed Pharmacother 109:2203–2209

Ziqing C, Yang Y, Liu LL, Lundqvist A (2019) Strategies to augment natural killer (NK) cell activity against solid tumors. Cancers 11:1040. https://doi.org/10.3390/cancers11071040

Immunology and Microbes

2

2.1 Immunology and Microbes

2.1.1 Immunology and Flora

Microbes living in our bodies, known as commensal microbes, have significant impacts on host health (Meng et al. 2019). At physiological conditions, microbes grow at various sites of the body such as oral cavities, respiratory tracts, and digestive tract. It has been estimated that the gastrointestinal (GI) tract may contain most colonizing microbes, about 100 trillion microbes, with hundreds of dominant bacterial species (Turner 2009; Qin et al. 2010; Blumberg and Powrie 2012).

The research problems relating interactions between gut microbiota and host health have been addressed since 1900. The idea was thought to be associated with the introduction of antibiotics, because it was thought that antibiotic is the magic solution for bacterial infections. Later, it was obvious that antibiotics complicated the detection options of microorganisms (Eckburg et al. 2005). Molecular detection techniques to increase the detection of microbes have been introduced since the 1970s (Sanger et al. 1977).

2.1.2 Immunology and Pathogens

Several studies have pointed out to the role of microbes and metabolites in both health and disease. As an example, induction of changes in the structure of gut microbiota has direct influences on human diseases, while induction of imbalance of gut microbiota (dysbiosis) causes changes of different metabolites (Clemente et al. 2012; Holmes et al. 2012; Sharon et al. 2014). Gut microbiota can make numerous metabolites including short-chain fatty acids (SCFAs), bile acids (BAs), and polysaccharide A (PSA) (Nicholson et al. 2012). SCFAs are considered a very important

A. J. Alkhatib, *The Role of Microbes in Autoimmune Diseases*,
https://doi.org/10.1007/978-981-19-1162-0_2

product of intestinal microbial metabolites because of the important role of each of acetic acid, butyric acid, and propionic acid (Koh et al. 2016).

It has been indicated by research studies that gut microbiota and metabolites can participate in the induction of several diseases such as autoimmune diseases (AIDs) (Cho and Blaser 2012; Sharon et al. 2014). Furthermore, it has been found that microbes have important influences on immune responses of the host (Brestoff and Artis 2013; Mu et al. 2017). It is believed that microbes induce their effects on immune response through the actions of regulatory T cell (Treg) and T helper cell 17 (Th17). Treg exhibits protective roles in autoimmune disease through its activation by intestinal bacteria such as Clostridium and Bacteroides species (Atarashi et al. 2013). The activation of Treg can be induced by signaling pathways resulting from bacterial metabolites or their structural molecules (Round and Mazmanian 2010; Arpaia et al. 2013; Smith et al. 2013; Tanoue et al. 2016). Th17 is more likely to be encountered in the intestine and works to maintain physiological immune status through attacking bacteria and removing pathogens (Huber et al. 2012). Moreover, environmental factors have important role on gut microbes, even more than host genes (Rothschild et al. 2018). New promising therapeutic options through transplantation of fecal microbiota have been reported (Mattila et al. 2012).

2.1.3 Microbiology and Immunological Regulatory Mechanisms

It seems that host and microbes have developed various regulatory mechanism to retain physiological conditions optimally for both. As an example, gut microbiota is a signaling molecule that has the direct ability to control harmonious equilibrium of intestinal mucosa. This regulatory mechanism is mediated through the effect of segmented filamentous bacterium (SFB) that induces the differentiation of Th17. This process is further facilitated by the existence of serum amyloid A (SAA). The process is further complicated by the presence of intestinal macrophages (CX3CR1, Mfs) and the expression of major histocompatibility complex (MHCII) on dendritic cells (DCs) to induce Th17 (Ivanov et al. 2009; Goto et al. 2014; Panea et al. 2015) (Fig. 2.1).

At the same time, Interleukin (IL)-17R-dependent signaling controls the growth of SFB through the action of various molecules such as α-defensins, polymeric immunoglobulin receptor, and Nox1 expression (Kumar et al. 2016).

It has been indicated that the homeostasis of intestinal mucosa can be kept through the induction of anti-inflammatory cytokine IL-10 release, which, in turn, can result from the growth of Clostridium strains in gut lamina propria (LP). This process also involves the expression of transforming growth factor–β (TGF-β), Foxp3+ Treg accumulations, which are facilitated by Clostridium strains (Atarashi et al. 2011). A relationship between periodontal disease (PD) and rheumatoid arthritis (RA) has been reported (de Pablo et al. 2008; Chen et al. 2013). It is believed that *Porphyromonas gingivalis (P. gingivalis)* is the link between PD and RA (Quirke et al. 2014). *P. gingivalis* is the main etiologic agent of PD and is merely the gram-negative bacteria able to release peptidylarginine deiminase (PAD)

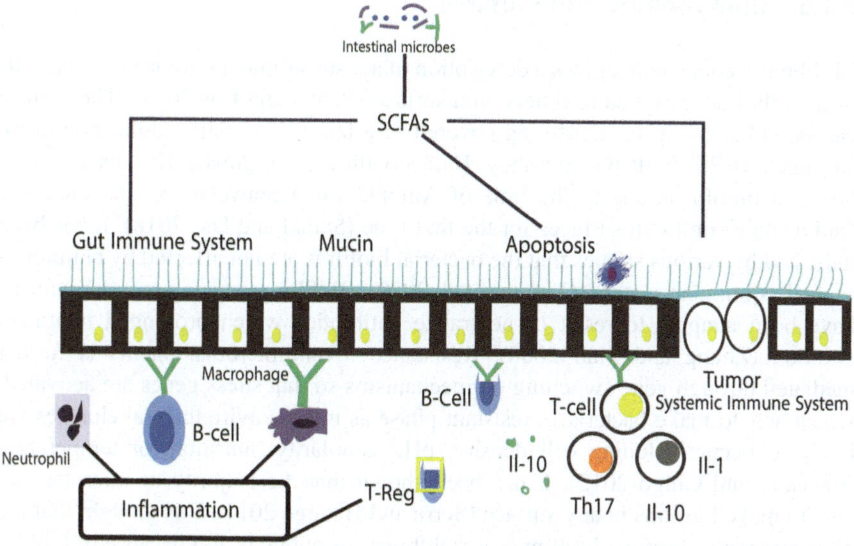

Fig. 2.1 The importance of short-chain fatty acids (SCFAs) in gut and systemic immune control

(McGraw et al. 1999). Anionic lipopolysaccharide (A-LPS) contributes in creating outer membrane vesicles (OMVs) through deacylation. *P. gingivalis* virulence factors are transported to synovium joint through OMVs (Veith et al. 2014). Within the same context, it has been found that OMVs can resist the immune system and worsen inflammatory conditions (Gui et al. 2016). Upon transporting of PAD, it can catalyze the transformation of arginine into citrulline, which increases its levels in synovium (Vossenaar et al. 2003; Maresz et al. 2013). Citrullinated proteins contribute to formation of autoimmunity through interacting with anti-cyclic citrullinated peptide antibody, RA specific indicator making immune complexes and binding Fc and C5a receptors in inflammatory cells, which, in turn, develops inflammatory conditions in both gums and synovium (Schellekens et al. 2000; Rosenstein et al. 2004; Avouac et al. 2006). Furthermore, *P. gingivalis* heat shock protein 60 has been found to participate in periodontitis-related AIDs, although the mechanism is unknown (Jeong et al. 2012; Kwon et al. 2016).

2.1.4 Signal Pathway of Microbial Dietary Metabolites

Metabolites resulting from microbial metabolism can induce pathologic changes following certain pathways such as the production of SCFAs and indole derivatives. SCFAs exert two functions through keeping the epithelial uprightness and exhibiting great anti-inflammatory reactions through adjusting immune response (Cummings 1983; Macia et al. 2012). SCFAs can prevent the production of inflammatory cytokines, macrophage activation, in addition to expression of Toll-like receptor 4 (TLR 4), and lowering the translocation of LPS (Liu et al. 2014; Qiao et al. 2014; Haghikia et al. 2015).

2.1.5 Immunology and Biofilms

Biofilm is a concept that gives a description of a distribution of microbes so that cells of microbes are attached to others on a surface (Samal and Das 2018). The surface can be either living or non-living covered in a templet of extracellular polymeric substance (EPS) built by microbes (Hall-Stoodley et al. 2004). The discovery of bacterial biofilm is due to the time of Antonie van Leeuwenhoek who observed "animalcule" on tooth surfaces for the first time (Samal and Das 2018). It has been indicated by various studies that the bacterial biofilms are not affected by antibiotics as well as immune system (Olivares et al. 2020). Biofilm producing microorganisms have been adapted to resist or neutralize antibiotics which prolonged treatment duration (Zhang and Mah 2008). Resistance to antimicrobial agents is further mediated through gene switching on mechanisms so that stress genes are activated, which acts to make bacteria in resistant phase as micro-environmental changes are likely to occur including cell density, pH, osmolarity, nutrition, or temperature (Lebeaux and Ghigo 2012). It has been shown that the majority of microbes are able to make biofilms in any surface (Serra and Hengge 2014). It is thought that the phenomenon of forming biofilm as a real threat for public health because of biofilm related diseases or its exhibited resistance for numerous antibiotics (Khan et al. 2014; Srivastava and Bhargava 2016; Jamal et al. 2018). The biofilm is designed up to minimize the exposure to immune cells such as white blood cells, through forming exopolymer (Thurlow et al. 2011). Bacterial biofilm also impedes the movement of white cells within biofilms as well as lowering the ability of white cells to exert their functions properly to produce reactive oxygen species and preventing bacterial phagocytosis (Bayer et al. 1991; Malic et al. 2011; Hobley et al. 2015; Flemming et al. 2016). Another important aspect related to biofilm is its ability to spread infection and cause septicemia through growth on implanted medical devices such as catheters used in different parts of the body (Wolcott et al. 2010). Biofilm is associated with dental caries, lung infections, and chronic wounds (Ciofu et al. 2015).

It is the biofilm size that is considered as critical in phagocytosis process. It has also been reported that among immunocompetent individuals, the activity of immune system is not effective against biofilms (Olivares et al. 2020). The phase of innate immune reaction implies the activation of neutrophils and macrophages after direct contact with bacteria. Then, neutrophils are accumulated close to the biofilm, this is accompanied with oxygen deficiency due to high active metabolism of oxidative metabolism because of the reduction of molecular oxygen in superoxide (Watters et al. 2016). The phagocytic cells try to get access to the extracellular matrix with difficulty. Their movement becomes more slow and phagocytes are influenced by bacterial enzymes. Moreover, the lysing process of neutrophils leads to disposition of toxic compound in the medium leading to tissue damage (Frangogiannis 2019). Another important aspect of biofilm actions against immune cells has been described as a neutralizing effect of specific antibodies against bacterial compounds produced by immune cells including the elastase, the LPS, or the flagellum. This implies that the memory function of immune reactions is disabled, even partly. What makes matters worse is the involvement of these antibodies in immune complexes formation, which, in turn, precipitate into host tissues such as lung parenchyma. This process will be followed by activation of complement system and activation of opsonization of neutrophils. The whole process leads to tissue degradation at interaction sites (Jensen et al. 2010).

2.1.6 The Role of Viral Hepatitis, EBV, and CMV in Autoimmunity

Liver is affected by autoimmune disease and autoimmune hepatitis (AIH) is developed. AIH is a severe inflammatory disease leading to breaking down of liver parenchyma by the action of autoimmune responses and fibrosis (Christen and Hintermann 2018; Mieli-Vergani et al. 2018; Webb et al. 2018). The incidence of AIH is 2 per 100,000 persons, and its prevalence is 15 per 100,000 people at global level (Christen and Hintermann 2018). AIH is likely to affect females more than males and individuals at any age (Manns et al. 2010). AIH is characterized by being hepatitis, existence of autoantibodies against liver antigens, in addition to increased liver enzymes (Muratori et al. 2016, 2018). There are other immune liver diseases including primary biliary cholangitis (PBC) and primary sclerosing cholangitis (PSC) which further complicated reaching a definite diagnosis of AIH (Boberg et al. 2011).

HCV infection is likely to participate as an etiologic agent in the occurrence of AIH (Christen and Hintermann 2018).

Microbes have roles in the development of several autoimmune diseases, and these microbes initiate both innate and adaptive immune responses. According to the type of microbe, the immune response can be presented as type 1 phenotype, in case of encountering intracellular pathogen including viruses, or type 2 phenotype (allergic), as encountered in removing extracellular parasites including helminthes (Christen and Hintermann 2018; Thakur et al. 2019).

According to this context, viruses are considered the chief factors involved in the initiation of autoimmune reactions that are mainly characterized by type 1 immune responses.

Manfredo Vieira et al. (2018) showed that the translocation of the gut microbes, *Enterococcus gallinarum*, to the liver activated autoimmunity responses in genetically modified mice to develop systemic lupus erythematosus (SLE) through the stimulated production of several mediators such as cytokines, autoantigens, and endogenous retrovirus proteins. It is worth mentioning that it may be required to have more than one pathogen to elucidate an autoimmune disease (Ercolini and Miller 2008). Maizels et al. (2014) reported that infection by helminthes is likely to offer protection against immune reactions. Furthermore, it is thought that helminthes have direct impacts on immune reactions against autoantigens in autoimmune diseases (Manfredo Vieira et al. 2018).

Researchers have proposed various mechanisms to explain the development of autoimmune responses. One of the first mechanisms rotates round the access of intracellular pathogens such as viruses. Viruses breakdown infected cells and increase the expression of MHC molecules on the membranes of infected cells, in addition to efficient action of antigen-presenting cells (APC) (Roche and Furuta 2015). According to this context, host epitopes are likely to be represented in a form that activates self T cells (Besser et al. 2019). The second proposed mechanism implies that infection causing pathogens activate the general host defense mechanisms to remove invading microbes (Kobayashi et al. 2018). The resulting inflammatory mechanisms include increased production of prostaglandins, cytokines, and chemokines, in addition to activation of the innate immunity and adaptive immune system (Nauseef 2007). Due to the consideration that, adaptive immune system specifically targets pathogens, other inflammatory mediators are expected to help in developing autoimmune reactions (von Herrath et al. 2003). The third mechanism considers the activation of T cell via interacting with superantigens which can bind MHC and TCR in an independent manner leading to reactions against self-component (Van Kaer 2018). The fourth mechanism is based on having some similar structural components between pathogens and host that lead to specific immune reactions which are likely to interact with host structures. This may lead to damage of the idea of self-tolerance (Cusick et al. 2012; Christen 2014) (Fig. 2.2).

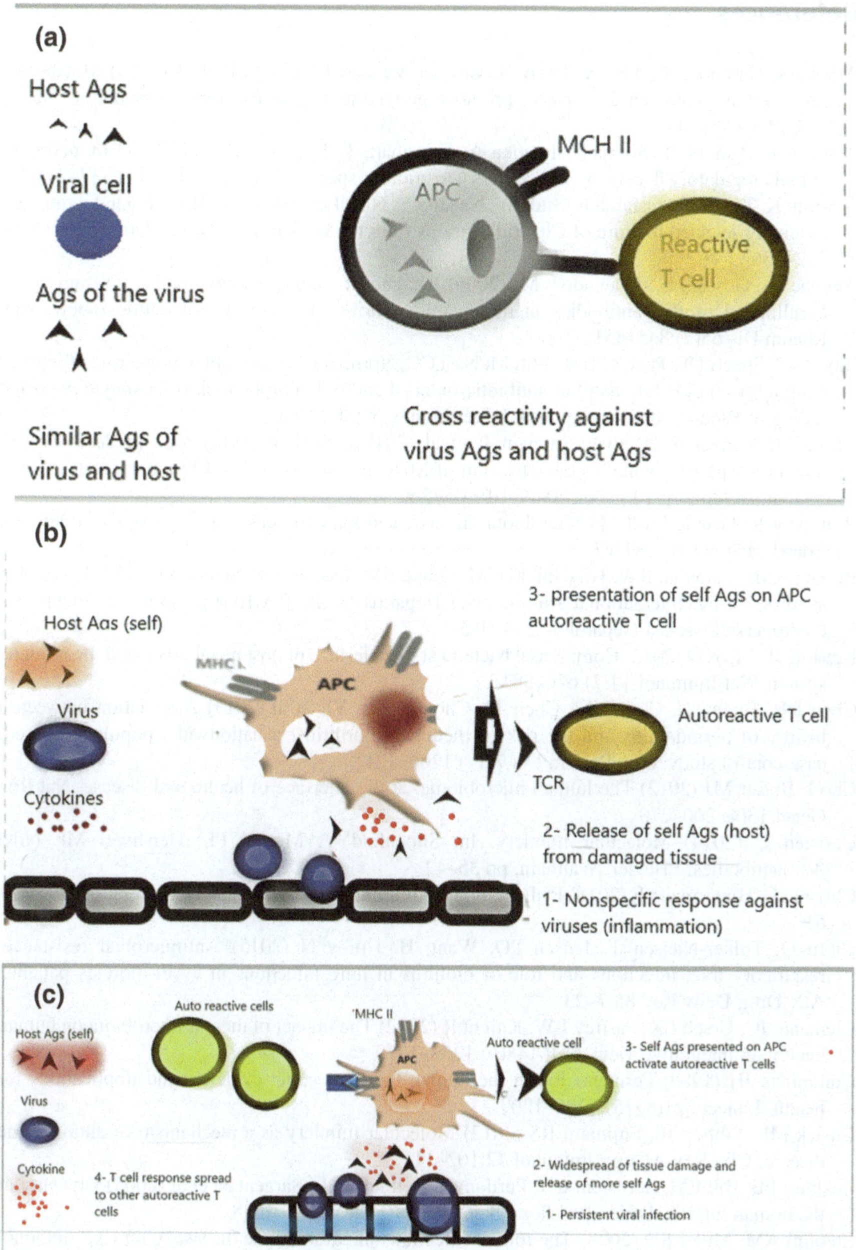

Fig. 2.2 (**a–c**) The role of viruses in the induction of autoimmunity. (**a**) Mimicking of molecules. (**b**) Bystander activation. (**c**) Spreading of epitope

References

Arpaia N, Campbell C, Fan X, Dikiy S, van der Veeken J, deRoos P et al (2013) Metabolites produced by commensal bacteria promote peripheral regulatory T-cell generation. Nature 504(7480):451–455

Atarashi K, Tanoue T, Shima T, Imaoka A, Kuwahara T, Momose Y et al (2011) Induction of colonic regulatory T cells by indigenous Clostridium species. Science 331(6015):337–341

Atarashi K, Tanoue T, Oshima K, Suda W, Nagano Y, Nishikawa H et al (2013) Treg induction by a rationally selected mixture of Clostridia strains from the human microbiota. Nature 500(7461): 232–236

Avouac J, Gossec L, Dougados M (2006) Diagnostic and predictive value of anti-cyclic citrullinated protein antibodies in rheumatoid arthritis: a systematic literature review. Ann Rheum Dis 65(7):845–851

Bayer AS, Speert DP, Park S, Tu J, Witt M, Nast CC, Norman DC (1991) Functional role of mucoid exopolysaccharide (alginate) in antibiotic-induced and polymorphonuclear leukocyte mediated killing of Pseudomonas aeruginosa. Infect Immun 59:302–308

Besser H, Yunger S, Merhavi-Shoham E et al (2019) Level of neo-epitope predecessor and mutation type determine T cell activation of MHC binding peptides. J Immunother Cancer 7: 135. https://doi.org/10.1186/s40425-019-0595-z

Blumberg R, Powrie F (2012) Microbiota, disease, and back to health: a metastable journey. Sci Transl Med 4(137):137rv7

Boberg KM, Chapman RW, Hirschfield GM, Lohse AW, Manns MP, Schrumpf E (2011) Overlap syndromes: the International Autoimmune Hepatitis Group (IAIHG) position statement on a controversial issue. J Hepatol 54:374–385

Brestoff JR, Artis D (2013) Commensal bacteria at the interface of host metabolism and the immune system. Nat Immunol 14(7):676–684

Chen HH, Huang N, Chen YM, Chen TJ, Chou P, Lee YL et al (2013) Association between a history of periodontitis and the risk of rheumatoid arthritis: a nationwide, population-based, case-control study. Ann Rheum Dis 72(7):1206–1211

Cho I, Blaser MJ (2012) The human microbiome: at the interface of health and disease. Nat Rev Genet 13(4):260–270

Christen U (2014) Molecular mimicry. In: Shoenfeld Y, Meroni PL, Gershwin ME (eds) Autoantibodies. Elsevier, Waltham, pp 35–42

Christen U, Hintermann E (2018) Pathogens and autoimmune hepatitis. Clin Exp Immunol 195:35–51

Ciofu O, Tolker-Nielsen T, Jensen PO, Wang H, Hoiby N (2015) Antimicrobial resistance, respiratory tract infections and role of biofilms in lung infections in cystic fibrosis patients. Adv Drug Deliv Rev 85:7–23

Clemente JC, Ursell LK, Parfrey LW, Knight R (2012) The impact of the gut microbiota on human health: an integrative view. Cell 148(6):1258–1270

Cummings JH (1983) Fermentation in the human large intestine: evidence and implications for health. Lancet 321(8335):1206–1209

Cusick MF, Libbey JE, Fujinami RS (2012) Molecular mimicry as a mechanism of autoimmune disease. Clin Rev Allergy Immunol 42:102–111

Eckburg PB, Bik EM, Bernstein CN, Purdom E, Dethlefsen L, Sargent M et al (2005) Diversity of the human intestinal microbial flora. Science 308(5728):1635–1638

Ercolini AM, Miller SD (2008) The role of infections in autoimmune disease. Clin Exp Immunol 155:1–15

Flemming H-C et al (2016) Biofilms: an emergent form of bacterial life. Nat Rev Microbiol 14:563–575

Frangogiannis NG (2019) The extracellular matrix in ischemic and nonischemic heart failure. Circ Res 125:117–146. https://doi.org/10.1161/CIRCRESAHA.119.311148

Goto Y, Panea C, Nakato G, Cebula A, Lee C, Diez MG et al (2014) Segmented filamentous bacteria antigens presented by intestinal dendritic cells drive mucosal Th17 cell differentiation. Immunity 40(4):594–607

Gui MJ, Dashper SG, Slakeski N, Chen YY, Reynolds EC (2016) Spheres of influence: Porphyromonas gingivalis outer membrane vesicles. Mol Oral Microbiol 31(5):365–378

Haghikia A, Jörg S, Duscha A, Berg J, Manzel A, Waschbisch A et al (2015) Dietary fatty acids directly impact central nervous system autoimmunity via the small intestine. Immunity 43(4): 817–829

Hall-Stoodley L, Costerton JW, Stoodley P (2004) Bacterial biofilms: from the natural environment to infectious diseases. Nat Rev Microbiol 2:95–108

von Herrath MG, Fujinami RS, Whitton LJ (2003) Microorganisms and autoimmunity: making the barren field fertile. Nat Rev Microbiol 1:151–157

Hobley L, Harkins C, MacPhee CE, Stanley-Wall NR (2015) Giving structure to the biofilm matrix: an overview of individual strategies and emerging common themes. FEMS Microbiol Rev 39: 649–669

Holmes E, Li JV, Marchesi JR, Nicholson JK (2012) Gut microbiota composition and activity in relation to host metabolic phenotype and disease risk. Cell Metab 16(5):559–564

Huber S, Gagliani N, Flavell RA (2012) Life, death, and miracles: Th17 cells in the intestine. Eur J Immunol 42(9):2238–2245

Ivanov II, Atarashi K, Manel N, Brodie EL, Shima T, Karaoz U et al (2009) Induction of intestinal Th17 cells by segmented filamentous bacteria. Cell 139(3):485–498

Jamal M, Ahmad W, Andleeb S, Jalil F, Imran M, Nawaz MA et al (2018) Bacterial biofilm and associated infections. J Chin Med Assoc 81:7–11. https://doi.org/10.1016/j.jcma.2017.07.012

Jensen PØ, Givskov M, Bjarnsholt T, Moser C (2010) The immune system vs. Pseudomonas aeruginosa biofilms. FEMS Immunol Med Microbiol 59:292–305. https://doi.org/10.1111/j.1574-695X.2010.00706.x

Jeong EJ, Lee JY, Kim SJ, Choi J (2012) Predominant immunoreactivity of Porphyromonas gingivalis heat shock protein in autoimmune diseases. J Periodontal Res 47(6):811–816

Khan S, Singh P, Ansari M, Asthana A (2014) Isolation of Shigella species and their resistance patterns to a panel of fifteen antibiotics in mid and far western region of Nepal. Asian Pac J Trop Dis 4:30–34

Kobayashi SD, Malachowa N, DeLeo FR (2018) Neutrophils and bacterial immune evasion. J Innate Immun 10(5–6):432–441. https://doi.org/10.1159/000487756

Koh A, De Vadder F, Kovatcheva-Datchary P, Bäckhed F (2016) From dietary fiber to host physiology: short-chain fatty acids as key bacterial metabolites. Cell 165(6):1332–1345

Kumar P, Monin L, Castillo P, Elsegeiny W, Horne W, Eddens T et al (2016) Intestinal Interleukin-17 receptor signaling mediates reciprocal control of the gut microbiota and autoimmune inflammation. Immunity 4(3):659–671

Kwon EY, Cha GS, Jeong E, Lee JY, Kim SJ, Surh CD et al (2016) Pep19 drives epitope spreading in periodontitis and periodontitis-associated autoimmune diseases. J Periodontal Res 51(3): 381–394

Lebeaux D, Ghigo J-M (2012) Infections associées aux biofilms - Quelles perspectives thérapeutiques issues de la recherche fondamentale?? Med Sci 28:727–739. https://doi.org/10.1051/medsci/2012288015

Liu B, Qian J, Wang Q, Wang F, Ma Z, Qiao Y (2014) Butyrate protects rat liver against total hepatic ischemia reperfusion injury with bowel congestion. PLoS One 9(8):e106184

Macia L, Thorburn AN, Binge LC, Marino E, Rogers KE, Maslowski KM et al (2012) Microbial influences on epithelial integrity and immune function as a basis for inflammatory diseases. Immunol Rev 245(1):164–176

Maizels RM, McSorley HJ, Smyth DJ (2014) Helminths in the hygiene hypothesis: sooner or later? Clin Exp Immunol 177:38–34

Malic S, Hill KE, Playle R, Thomas DW, Williams DW (2011) In vitro interactions of chronic wound bacteria in biofilms. J Wound Care 20:569–577

Manfredo Vieira S, Hiltensperger M, Kumar V et al (2018) Translocation of a gut pathobiont drives autoimmunity in mice and humans. Science 359:1156–1161

Manns MP, Czaja AJ, Gorham JD et al (2010) Diagnosis and management of autoimmune hepatitis. Hepatology 51:2193–2213

Maresz KJ, Hellvard A, Sroka A, Adamowicz K, Bielecka E, Koziel J et al (2013) Porphyromonas gingivalis facilitates the development and progression of destructive arthritis through its unique bacterial peptidylarginine deiminase (PAD). PLoS Pathog 9(9):e1003627

Mattila E, Uusitalo-Seppälä R, Wuorela M, Lehtola L, Nurmi H, Ristikankare M et al (2012) Fecal transplantation, through colonoscopy, is effective therapy for recurrent Clostridium difficile infection. Gastroenterology 142(3):490–496

McGraw WT, Potempa J, Farley D, Travis J (1999) Purification, characterization, and sequence analysis of a potential virulence factor from Porphyromonas gingivalis, peptidylarginine deiminase. Infect Immun 67(7):3248–3256

Meng X, Zhou H-Y, Shen H-H, Lufump E, Li X-M, Guo B, Li B-Z (2019) Microbe-metabolite-host axis, two-way action in the pathogenesis and treatment of human autoimmunity. Autoimmun Rev 18:455–475

Mieli-Vergani G, Vergani D, Czaja AJ et al (2018) Autoimmune hepatitis. Nat Rev Dis Primers 4: 18017

Mu Q, Zhang H, Liao X, Lin K, Liu H, Edwards MR et al (2017) Control of lupus nephritis by changes of gut microbiota. Microbiome 5(1):73

Muratori P, Lalanne C, Barbato E et al (2016) Features and progression of asymptomatic autoimmune hepatitis in Italy. Clin Gastroenterol Hepatol 14:139–146

Muratori P, Carbone M, Stangos G et al (2018) Clinical and prognostic implications of acute onset of autoimmune hepatitis: an Italian multicentre study. Dig Liver Dis 50:698–702

Nauseef WM (2007) How human neutrophils kill and degrade microbes: an integrated view. Immunol Rev 219:88–102

Nicholson JK, Holmes E, Kinross J, Burcelin R, Gibson G, Jia W et al (2012) Host-gut microbiota metabolic interactions. Science 336(6086):1262–1267

Olivares E, Badel-Berchoux S, Provot C, Prévost G, Bernardi T, Jehl F (2020) Clinical impact of antibiotics for the treatment of Pseudomonas aeruginosa biofilm infections. Front Microbiol 10: 2894. https://doi.org/10.3389/fmicb.2019.02894

de Pablo P, Dietrich T, McAlindon TE (2008) Association of periodontal disease and tooth loss with rheumatoid arthritis in the US population. J Rheumatol 35(1):70–76

Panea C, Farkas AM, Goto Y, Abdollahi-Roodsaz S, Lee C, Koscsó B et al (2015) Intestinal monocyte-derived macrophages control commensal-specific Th17 responses. Cell Rep 12(8): 1314–1324

Qiao YL, Qian JM, Wang FR, Ma ZY, Wang QW (2014) Butyrate protects liver against ischemia reperfusion injury by inhibiting nuclear factor kappa B activation in Kupffer cells. J Surg Res 187(2):653–659

Qin J, Li R, Raes J, Arumugam M, Burgdorf KS, Manichanh C et al (2010) A human gut microbial gene catalogue established by metagenomic sequencing. Nature 464(7285):59–65

Quirke AM, Lugli EB, Wegner N, Hamilton BC, Charles P, Chowdhury M et al (2014) Heightened immune response to autocitrullinated Porphyromonas gingivalis peptidylarginine deiminase: a potential mechanism for breaching immunologic tolerance in rheumatoid arthritis. Ann Rheum Dis 73(1):263–269

Roche PA, Furuta K (2015) The ins and outs of MHC class II-mediated antigen processing and presentation. Nat Rev Immunol 15(4):203–216

Rosenstein ED, Greenwald RA, Kushner LJ, Weissmann G (2004) Hypothesis: the humoral immune response to oral bacteria provides a stimulus for the development of rheumatoid arthritis. Inflammation 28(6):311–318

Rothschild D, Weissbrod O, Barkan E, Kurilshikov A, Korem T, Zeevi D et al (2018) Environment dominates over host genetics in shaping human gut microbiota. Nature 555(7695):210–215

Round JL, Mazmanian SK (2010) Inducible Foxp3+ regulatory T-cell development by a commensal bacterium of the intestinal microbiota. Proc Natl Acad Sci U S A 107(27):12204–12209

Samal S, Das PK (2018) Microbial biofilms: pathogenicity and treatment strategies. PharmaTutor 6(1):16–22. https://doi.org/10.29161/PT.v6.i1.2017.16

Sanger F, Nicklen S, Coulson AR (1977) DNA sequencing with chain-terminating inhibitors. Proc Natl Acad Sci U S A 74(12):5463–5467

Schellekens GA, Visser H, de Jong BA, van den Hoogen FH, Hazes JM, Breedveld FC et al (2000) The diagnostic properties of rheumatoid arthritis antibodies recognizing a cyclic citrullinated peptide. Arthritis Rheum 43(1):155–163

Serra DO, Hengge R (2014) Stress responses go three dimensional - the spatial order of physiological differentiation in bacterial macrocolony biofilms. Environ Microbiol 16:1455–1471. https://doi.org/10.1111/1462-2920.12483

Sharon G, Garg N, Debelius J, Knight R, Dorrestein PC, Mazmanian SK (2014) Specialized metabolites from the microbiome in health and disease. Cell Metab 20(5):719–730

Smith PM, Howitt MR, Panikov N, Michaud M, Gallini CA, Bohlooly-Y M et al (2013) The microbial metabolites, short-chain fatty acids, regulate colonic Treg cell homeostasis. Science 341(6145):569–573

Srivastava S, Bhargava A (2016) Biofilms and human health. Biotechnol Lett 38:1–22. https://doi.org/10.1007/s10529-015-1960-8

Tanoue T, Atarashi K, Honda K (2016) Development and maintenance of intestinal regulatory T cells. Nat Rev Immunol 16(5):295–309

Thakur A, Mikkelsen H, Jungersen G (2019) Intracellular pathogens: host immunity and microbial persistence strategies. Hindawi J Immunol Res 2019:1356540, 24p. https://doi.org/10.1155/2019/1356540

Thurlow LR, Hanke ML, Fritz T, Angle A, Aldrich A, Williams SH, Engebretsen IL, Bayles KW, Horswill AR, Kielian T (2011) Staphylococcus aureus biofilms prevent macrophage phagocytosis and attenuate inflammation in vivo. J Immunol 186:6585–6596

Turner JR (2009) Intestinal mucosal barrier function in health and disease. Nat Rev Immunol 9(11): 799–809

Van Kaer L (2018) How superantigens bind MHC. J Immunol 201(7):1817–1818. https://doi.org/10.4049/jimmunol.1801104

Veith PD, Chen YY, Gorasia DG, Chen D, Glew MD, O'Brien-Simpson NM et al (2014) Porphyromonas gingivalis outer membrane vesicles exclusively contain outer membrane and periplasmic proteins and carry a cargo enriched with virulence factors. J Proteome Res 13(5): 2420–2432

Vossenaar ER, Zendman AJ, van Venrooij WJ, Pruijn GJ (2003) PAD, a growing family of citrullinating enzymes: genes, features and involvement in disease. BioEssays 25(11): 1106–1118

Watters C, Fleming D, Bishop D, Rumbaugh KP (2016) Host responses to biofilm. Prog Mol Biol Transl Sci 142:193–239. https://doi.org/10.1016/bs.pmbts.2016.05.007

Webb GJ, Hirschfield GM, Krawitt EL, Gershwin ME (2018) Cellular and molecular mechanisms of autoimmune hepatitis. Annu Rev Pathol 13:247–292

Wolcott RD, Rumbaugh KP, James G, Schultz G, Phillips P, Yang Q, Watters C, Steward PS, Dowd SE (2010) Biofilm maturity studies indicate sharp debridement opens a time-dependent therapeutic window. J Wound Care 19:320–328

Zhang L, Mah T-F (2008) Involvement of a novel efflux system in biofilmspecific resistance to antibiotics. J Bacteriol 190:4447–4452. https://doi.org/10.1128/JB.01655-07

3 Types of Hypersensitivities (Updates)

3.1 Introduction

This chapter gives recent updates regarding types of hypersensitivities. Hypersensitivity is also known as hypersensitivity reaction, or intolerance. It implies unpleasant reactions played by immune system (Doña et al. 2020).

3.2 Type I Hypersensitivity

Type I hypersensitivity is considered as an allergic reaction resulting from the re-exposure to certain antigen, called allergen reactions that can be local or systemic (Godwin and Crane 2020). Type I hypersensitivity symptoms are varied from tolerable irritation to death due to anaphylactic shock (Anto et al. 2017; Valenta et al. 2018). Therapeutic options include the use of epinephrine, antihistamines, and corticosteroids (Wood 2016; Dona and Suphioglu 2020).

In hypersensitivity type I, the exposure to antigen can be mediated by various routes such as inhalation, ingestion, injection, or direct contact (Palmer et al. 2013; Brough et al. 2015; Inomata et al. 2018). Following the exposure to the antigen, the B lymphocytes, known as "naïve," become aware of the antigen, primed, and then they get differentiated as antibody-secreting cells and IgE antibody class (Justiz Vaillant et al. 2020). It is worth mentioning that the main difference between type I hypersensitivity response and regular immune response is that the class of immunoglobulin secreted by plasma cells in type I hypersensitivity is IgE, while in regular response is mainly IgG or IgM (Hofmaier et al. 2015; Saunders et al. 2019). IgE can specifically bind to Fc receptors on tissue mast cells and blood basophils (Sutton and Davies 2015; Saunders et al. 2019). At this stage, mast cells and basophils are sensitized as they have been coated by IgE (Fig. 3.1). Future exposure to the same allergen makes sensitized cells to be degranulated causing the secretion of histamine, leukotriene, and prostaglandins, which in turn impacts the surrounding tissues

A. J. Alkhatib, *The Role of Microbes in Autoimmune Diseases*,
https://doi.org/10.1007/978-981-19-1162-0_3

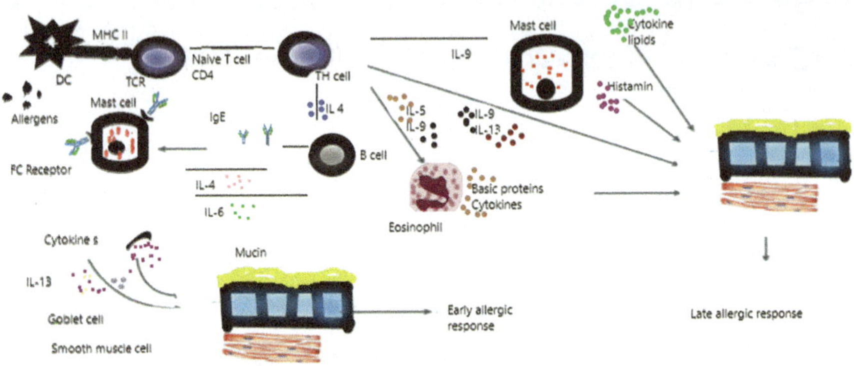

Fig. 3.1 Schematic diagram of hypersensitivity type I

(Butterfield and Weiler 2008; Theoharides et al. 2019; Gasser et al. 2020). These products mainly work in the vasodilation and smooth muscle contraction (Hedqvist et al. 2000). Examples of type I hypersensitivity include, but not restricted to, allergic asthma, allergic rhinitis (hay fever), and allergic conjunctivitis (Theoharides et al. 2015, 2019).

3.3 Type II Hypersensitivity: Antibody-Dependent

Type II hypersensitivity is characterized by reactions leading to the production of antibodies, usually IgG and IgM, that interact with antigens recognized on the individual's cell components (Moro et al. 2010; Roan et al. 2019). Antigens can be intrinsic, self-antigen, or extrinsic resulting from exposure to foreign antigens (bio.libretexts.org 2020). IgG and IgM antibodies interact with the antigens making immune complexes that work to activate the classical pathways of complement system to remove cells that are recognized as foreign antigen presenting cells (Justiz Vaillant et al. 2020). Mediators resulting from acute inflammation are accumulated at the area leading to the attack of membranes by complexes which, in turn, induce cellular death (Vries et al. 2017). Cellular phagocytosis is facilitated by the cells that express Fc receptors. These Fc receptors identify the surface-bound antibody and the complement receptors, which recognize the protein complement bound to the surface. In addition to previous considerations, these cells can be identified by both macrophages or dendritic cells; the cells that are acting present antigens leading to provoking of a B-cell response in which antibodies are produced against the foreign antigen. As an illustrating example, the reaction to penicillin, as the drug can bind to red blood cells and cause them to recognize each other as different. B-cell proliferation will occur and antibodies against the drug will be produced. Another form of type II hypersensitivity is called antibody-dependent cytotoxicity (ADCC). Here, cells that display foreign antigens are distinguished by antibodies (IgG or IgM). Here, cells that display the foreign antigen are labeled with antibodies (IgG or

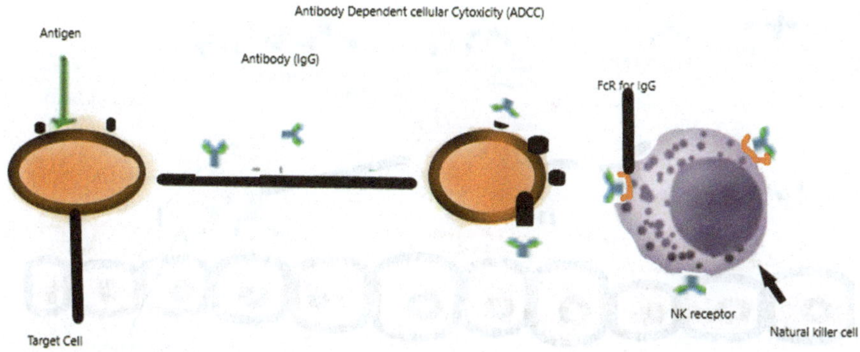

Fig. 3.2 Schematic diagram of hypersensitivity type II

IgM). These labeled cells are then recognized by the cells of the natural killer (NK) and macrophages (they are recognized by the IgG bound to the cell surface receptor, CD16 (Fc? RIII)), which in turn kills these labeled cells. Some examples: autoimmune hemolytic anemia, Goodpasture syndrome, and erythroblastosis fetal (Edberg et al. 1998; Ravetch 2003; van Lookeren et al. 2007; Bakema and van Egmond 2011; Rosales and Uribe-Querol 2013a, b; Uribe-Querol and Rosales 2020) (Fig. 3.2).

3.4 Type III: Immunological Complex

Type III hypersensitivity reactions are based on the existence of insoluble immune complexes resulting from the accretions of IgG and IgM antigens and antibodies that have been made in the blood and precipitated in several tissues including the skin, kidneys, and joints (Fan et al. 2020). Antibody deposition can induce an immune response according to the activation pathway of the classic supplement: to eliminate cells that contain foreign antigens (Hsiao et al. 2014).

The formation of complexes occurs in two stages: first, when IgG and IgM antibodies bind to an antigen, and then, when the complexes combine to form larger complexes that can be cleared from the body (Rojko et al. 2014). It is in the first stage of this formation that clearance is impossible, and the antibody and antigen complex spread as previously described (Usman and Annamaraju 2020). The reaction can take anywhere from hours to days to create (Fan et al. 2020). With the influx of phagocytes and granulocytes, as well as the release of inflammatory mediators, tissue damage occurs at the site of the immune complex (Usman and Annamaraju 2020). Some examples include glomerulonephritis with an immune complex, rheumatoid arthritis—a form of arthritis that affects the joints, sickness caused by serum, bacterial endocarditis, subacute malaria signs and symptoms, and systemic lupus erythematosus (SLE)—a form of lupus that affects the whole body (Patterson-Fortin et al. 2016; Fan et al. 2020; Rhee et al. 2020; Usman and Annamaraju 2020) (Fig. 3.3).

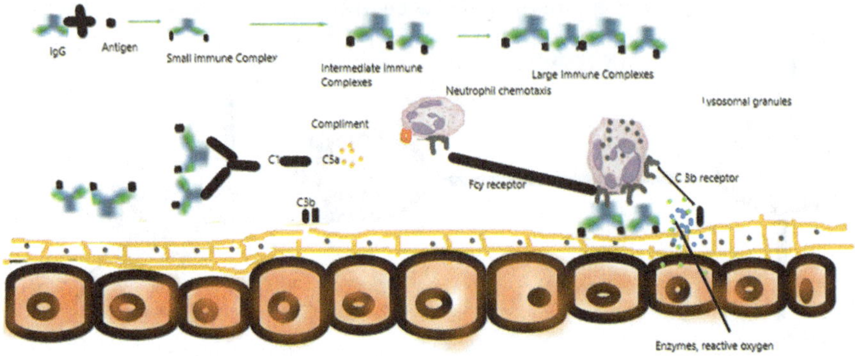

Fig. 3.3 Schematic diagram representing hypersensitivity type III (immune complexes)

3.5 Type IV Hypersensitivity Reactions: Cell-Mediated Reaction

A type IV hypersensitivity reaction is a cell-mediated reaction that may occur because of contact with some allergens, resulting in contact dermatitis, or because of certain diagnostic procedures, such as the tuberculin skin test. To control this disease, such allergens must be avoided. This activity goes into how to evaluate and treat type IV hypersensitivity reactions and emphasizes the importance of a multidisciplinary team in providing better treatment for patients with this condition (Marwa and Kondamudi 2020).

T cells cause an inflammatory response to exogenous or endogenous antigens in a type IV hypersensitivity reaction. Other cells, such as monocytes, eosinophils, and neutrophils, can be involved in certain cases. An initial local immune and inflammatory response occurs after antigen exposure, which attracts leukocytes. The antigen ingested by macrophages and monocytes is then presented to T cells, which become sensitized and activated as a result. These cells then produce cytokines and chemokines, which can destroy tissue and lead to illness. Contact dermatitis and drug hypersensitivity are two illnesses that may result from type IV hypersensitivity reactions. Based on the type of T cell (CD4 T helper type 1 and type 2 cells) involved and the cytokines/chemokines released, type IV reactions are further divided into type IVa, IVb, IVc, and IVd (Pichler 2003; Marwa and Kondamudi 2020).

Delayed hypersensitivity is essential for our body's ability to combat intracellular pathogens like mycobacteria and fungi. They're also essential for tumor immunity and transplant rejection. Patients with acquired immunodeficiency syndrome (AIDS) have a faulty type IV hypersensitivity reaction because their CD4 cell count decreases over time (Justiz Vaillant et al. 2020; Usman and Annamaraju 2020). Type IV hypersensitivity is a common affliction among those who are sensitive. The prevalence of touch hypersensitivity, for example, is estimated to be between 1% and 6% of the population. Drug allergy is a mild subtype of type IV hypersensitivity

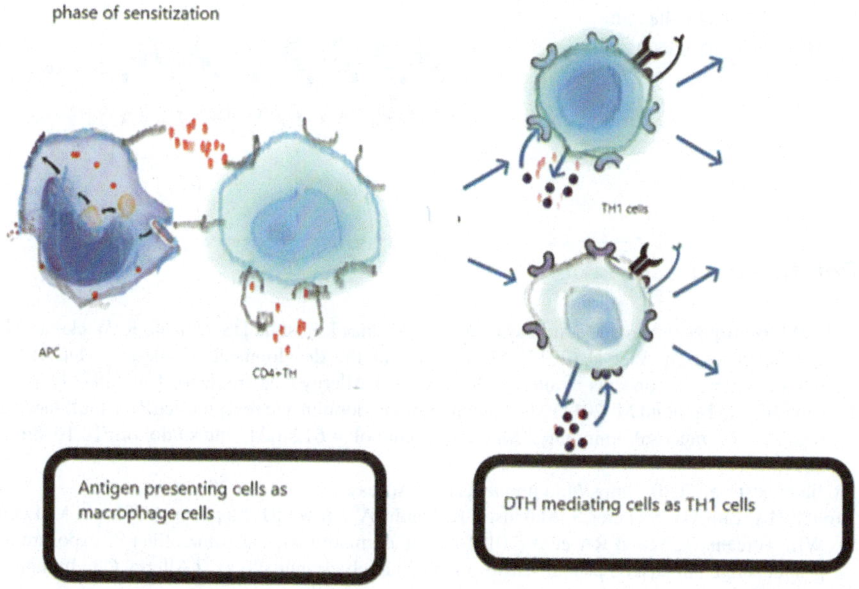

Fig. 3.4 Schematic diagram of hypersensitivity type IV

reactions. Drug allergy is often considered a separate condition and accounts for about one-seventh of all drug side effects. About 2% to 3% of hospitalized patients have allergic skin reactions. Epidemiological research in Western Europe has discovered a connection between atopy and the purified protein derivative (PPD) response in healthy people. PPD reactions of more than 3 mm are slightly more common in patients with atopy than in the general population, according to a major cross-sectional study conducted in Sweden (15.1% vs. 14.7%). PPD responses with a result greater than 10 mm, on the other hand, were found to be 1.4% in normal children and 1.2% in allergic children (Grüber and Paul 2002; Pichler 2003; Czarnobilska et al. 2007) (Fig. 3.4).

3.6 Type V Hypersensitivity Reactions

This is another form that is often used to distinguish type II (especially in the UK) (Rajan 2003). Instead of binding to cell surfaces, antibodies recognize and attach to cell surface receptors, which either prevents the intended ligand from binding to the receptor or mimics the ligand's effects, causing cell signaling to be disrupted (immunopaedia.org.za 2021) (Fig. 3.5).

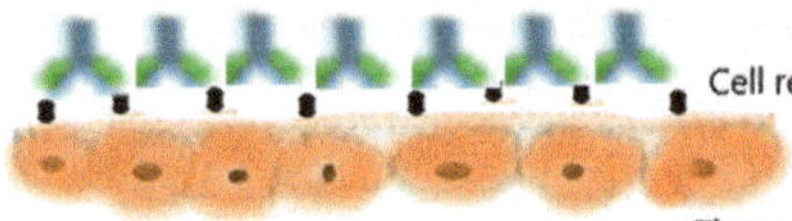

Fig. 3.5 Schematic diagram of hypersensitivity type V

References

Anto JM, Bousquet J, Akdis M, Auffray C, Keil T, Momas I, Postma DS, Valenta R, Wickman M, Cambon-Thomsen A et al (2017) Mechanisms of the development of allergy (MeDALL): introducing novel concepts in allergy phenotypes. J Allergy Clin Immunol 139:388–399

Bakema JE, van Egmond M (2011) The human immunoglobulin a fc receptor FcαRI: a multifaceted regulator of mucosal immunity. Mucosal Immunol 4:612–624. https://doi.org/10.1038/mi.2011.36

bio.libretexts.org (2020). https://bio.libretexts.org. Accessed 17 Aug 2020

Brough HA, Liu AH, Sicherer S, Makinson K, Douiri A, Brown SJ, Stephens AC, Irwin McLean WH, Turcanu V, Wood RA et al (2015) Atopic dermatitis increases the effect of exposure to peanut antigen in dust on peanut sensitization and likely peanut allergy. J Allergy Clin Immunol 135:1641–1670

Butterfield JH, Weiler CR (2008) Prevention of mast cell activation disorder-associated clinical sequelae of excessive prostaglandin D(2) production. Int Arch Allergy Immunol 147:338–343

Czarnobilska E, Obtułowicz K, Wsołek K (2007) Type IV of hypersensitivity and its subtypes. Przegl Lek 64(7–8):506–508

Dona DW, Suphioglu C (2020) Egg allergy: diagnosis and immunotherapy. Int J Mol Sci 21:5010. https://doi.org/10.3390/ijms21145010

Doña I, Pérez-Sánchez N, Eguiluz-Gracia I et al (2020) Progress in understanding hypersensitivity reactions to nonsteroidal anti-inflammatory drugs. Allergy 75:561–575. https://doi.org/10.1111/all.14032

Edberg JC, Moon JJ, Chang DJ, Kimberly RP (1998) Differential regulation of human neutrophil FcγRIIa (CD32) and FcγRIIIb (CD16)-induced Ca2+ transients. J Biol Chem 273:8071–8079. https://doi.org/10.1074/jbc.273.14.8071

Fan H, Zhang S, Li N, Fan P, Hu X, Liang K, Cheng X, Wu Y (2020) Stable expression ratios of five pyroptosis-inducing cytokines in the spleen and thymus of mice showed potential immune regulation at the organ level. Lupus 29(3):290–302

Gasser P, Tarchevskaya SS, Guntern P et al (2020) The mechanistic and functional profile of the therapeutic anti-IgE antibody ligelizumab differs from omalizumab. Nat Commun 11:165. https://doi.org/10.1038/s41467-019-13815-w

Godwin L, Crane JS (2020) Biochemistry, immunoglobulin E (IgE). In: StatPearls [Internet]. StatPearls, Treasure Island, FL. https://www.ncbi.nlm.nih.gov/books/NBK541058

Grüber C, Paul KP (2002) Tuberculin reactivity and allergy. Allergy 57(4):277–280

Hedqvist P, Gautam N, Lindbom L (2000) Interactions between leukotrienes and other inflammatory mediators/modulators in the microvasculature. Am J Respir Crit Care Med 161:117–119

Hofmaier S, Hatzler L, Rohrbach A, Panetta V, Hakimeh D, Bauer CP et al (2015) "Default" versus "pre-atopic" IgG responses to foodborne and airborne pathogenesis-related group 10 protein molecules in birch-sensitized and nonatopic children. J Allergy Clin Immunol 135:1367–74e1–8. https://doi.org/10.1016/j.jaci.2014.09.048

Hsiao YP, Tsai JD, Muo CH, Tsai CH, Sung FC, Liao YT, Chang YJ, Yang JH (2014) Atopic diseases and systemic lupus erythematosus: an epidemiological study of the risks and correlations. Int J Environ Res Public Health 11(8):8112–8122 immunopaedia.org.za (2021). https://www.immunopaedia.org.za/immunology/archive/type-i-iv-hypersensitivity-reactions/immune-complex-formation/hypersensitivity-reactions/?print=pdf. Accessed 10 Mar 2021

Inomata N, Kawano K, Aihara M (2018) Bird-egg syndrome induced by α-livetin sensitization in a budgerigar keeper: successful induction of tolerance by avoiding exposure to avians. Allergol Int 68(2):282–284

Justiz Vaillant AA, Vashisht R, Zito PM (2020) Immediate hypersensitivity reactions. In: StatPearls. StatPearls, Treasure Island, FL. https://www.ncbi.nlm.nih.gov/books/NBK513315/. Accessed 16 Aug 2020

Marwa K, Kondamudi NP (2020) Type IV hypersensitivity reaction. In: StatPearls [Internet]. StatPearls, Treasure Island, FL. https://www.ncbi.nlm.nih.gov/books/NBK562228. Accessed 10 Sep 2020

Moro K et al (2010) Innate production of T(H)2 cytokines by adipose tissue-associated c-kit (+) Sca-1(+) lymphoid cells. Nature 463(7280):540–544

Palmer DJ, Metcalfe J, Makrides M, Gold MS, Quinn P, West CE, Loh R, Prescott SL (2013) Early regular egg exposure in infants with eczema: a randomized controlled trial. J Allergy Clin Immunol 132:3873–3892

Patterson-Fortin J, Harris CM, Niranjan-Azadi A, Melia M (2016) Serum sickness-like reaction after the treatment of cellulitis with amoxicillin/clavulanate. BMJ Case Rep 2016

Pichler WJ (2003) Delayed drug hypersensitivity reactions. Ann Intern Med 139(8):683–693. https://doi.org/10.7326/0003-4819-139-8-200310210-00012. PMID: 14568857

Rajan TV (2003) The Gell–Coombs classification of hypersensitivity reactions: a re-interpretation. Trends Immunol 24(7):376–379. https://doi.org/10.1016/S1471-4906(03)00142-X

Ravetch JV (2003) Fc receptors. In: Paul WE (ed) Fundamental immunology. Lippincott Williams & Wilkins, Philadelphia, pp 631–684

Rhee C, Kadri SS, Dekker JP, Danner RL, Chen HC, Fram D, Zhang F, Wang R, Klompas M (2020) CDC prevention epicenters program. Prevalence of antibiotic-resistant pathogens in culture-proven sepsis and outcomes associated with inadequate and broad-Spectrum empiric antibiotic use. JAMA Netw Open 3(4):e202899

Roan F, Obata-Ninomiya K, Ziegler SF (2019) Epithelial cell-derived cytokines: more than just signaling the alarm. J Clin Invest 129(4):1441–1451. https://doi.org/10.1172/JCI124606

Rojko JL, Evans MG, Price SA et al (2014) Formation, clearance, deposition, pathogenicity, and identification of biopharmaceutical-related immune complexes: review and case studies. Toxicol Pathol 42(4):725–764. https://doi.org/10.1177/0192623314526475

Rosales C, Uribe-Querol E (2013a) Antibody—fc receptor interactions in antimicrobial functions. Curr Immunol Rev 9:44–55. https://doi.org/10.2174/1573395511309010006

Rosales C, Uribe-Querol E (2013b) Fc receptors: cell activators of antibody functions. Adv Biosci Biotechnol 4:21–33. https://doi.org/10.4236/abb.2013.44A004

Saunders SP, Ma EGM, Aranda CJ, Curotto de Lafaille MA (2019) Non-classical B cell memory of allergic IgE responses. Front Immunol 10:715. https://doi.org/10.3389/fimmu.2019.00715

Sutton BJ, Davies AM (2015) Structure and dynamics of IgE-receptor interactions: FcepsilonRI and CD23/FcepsilonRII. Immunol Rev 268:222–235. https://doi.org/10.1111/imr.12340

Theoharides TC, Valent P, Akin C (2015) Mast cells, mastocytosis, and related disorders. N Engl J Med 373:163–172

Theoharides TC, Tsilioni I, Ren H (2019) Recent advances in our understanding of mast cell activation – or should it be mast cell mediator disorders? Expert Rev Clin Immunol 15(6): 639–656. https://doi.org/10.1080/1744666X.2019.1596800

Uribe-Querol E, Rosales C (2020) Phagocytosis: our current understanding of a universal biological process. Front Immunol 11:1066. https://doi.org/10.3389/fimmu.2020.01066

Usman N, Annamaraju P (2020) Type III hypersensitivity reaction. In: StatPearls [Internet]. StatPearls, Treasure Island, FL. https://www.ncbi.nlm.nih.gov/books/NBK559122. Accessed 14 Dec 2020

Valenta R, Karaulov A, Niederberger V, Gattinger P, van Hage M, Flicker S, Linhart B, Campana R, Focke-Tejkl M, Curin M et al (2018) Chapter five—molecular aspects of allergens and allergy. In: Alt F (ed) Advances in immunology, vol 138. Academic, Cambridge, MA, pp 195–256

van Lookeren M, Campagne CW, Brown EJ (2007) Macrophage complement receptors and pathogen clearance. Cell Microbiol 9:2095–2102. https://doi.org/10.1111/j.1462-5822.2007.00981.x

Vries TB, Boerma S, Doornebal J, Dikkeschei B, Stegeman C, Veneman TF (2017) Goodpasture's syndrome with negative anti-glomerular basement membrane antibodies. Eur J Case Rep Intern Med 4(8):000687

Wood RA (2016) New horizons in allergen immunotherapy. JAMA 315:1711–1712

Autoimmunity 4

4.1 Introduction to Autoimmunity

Autoimmune diseases (such as type 1 diabetes (T1D), multiple sclerosis (MS), and systemic lupus erythematosus (SLE)) are among the most common chronic illnesses caused by a dysregulated inflammatory response against self-antigens. Autoimmunity affects about 5% of the general population, and the geoepidemiology of these diseases becomes more complicated when certain factors like age, gender, ethnicity, and other demographic characteristics are considered (Cooper et al. 2009; Alriyami and Polychronakos 2021).

Autoimmunity occurs when an organism fails to recognize its own constituent parts as self, allowing the immune system to attack its own cells and tissues (Rodriguez-Rodriguez 2015; Moulton et al. 2017). An autoimmune disease is described as a disease caused by an abnormal immune response (Moulton and Tsokos 2015). Celiac disease, diabetes mellitus type 1 (IDDM), sarcoidosis, systemic lupus erythematosus (SLE), Sjögren's syndrome, Churg-Strauss syndrome, Hashimoto's thyroiditis, Graves' disease, idiopathic thrombocytopenic purpura, Addison's disease, rheumatoid arthritis (RA), and allergies are several of the more common examples of autoimmune diseases (bio.libretexts.org 2021). Steroids are often used to treat autoimmune diseases (Velikova and Georgiev 2021).

It is not a new belief that an individual's immune system is completely incapable of identifying self-antigens. Paul Ehrlich suggested the idea of horror autotoxicus at the turn of the twentieth century, in which a "human" body does not mount an immune response to its own tissues. As a result, any autoimmune response was regarded as abnormal and associated with human disease. It is now widely recognized that autoimmune responses (also known as "natural autoimmunity") are a common part of vertebrate immune systems that are usually prevented from causing disease by the phenomenon of immunological tolerance to self-antigens. Autoimmunity is not to be confused with alloimmunity (Cui et al. 2010; Seetharam et al. 2010).

A. J. Alkhatib, *The Role of Microbes in Autoimmune Diseases*,
https://doi.org/10.1007/978-981-19-1162-0_4

An immune response against one's own body cells cannot be triggered under normal circumstances (Brodin 2021). Immune cells, on the other hand, occasionally make a mistake and attack the very cells they are supposed to protect (Davis and Hollis 2016).

While a high level of autoimmunity is harmful, a low level of autoimmunity can be advantageous (Zarrin et al. 2021). First, low-level autoimmunity can help CD8+ T cells in recognizing neoplastic cells, reducing the incidence of cancer (Mpakali and Stratikos 2021). Second, autoimmunity can play a role in enabling a rapid immune response in the early stages of an infection, when foreign antigen availability is limited (i.e., when there are few pathogens present) (Gunjan et al. 2018).

There are a variety of immunodeficiency syndromes with autoimmunity-like clinical and laboratory features (Marwaha et al. 2012). In these patients, the immune system's decreased ability to clear infections may be to blame for autoimmunity, which is caused by constant immune system activation (Halabi-Tawil et al. 2009). Multiple autoimmune disorders, such as inflammatory bowel disease, autoimmune thrombocytopenia, and autoimmune thyroid disease, are seen in people with common variable immunodeficiency (CVID) (Meffre 2011). Another example is familial hemophagocytic lymphohistiocytosis, an autosomal recessive primary immunodeficiency (Kusne et al. 2021).

Many autoimmune diseases, such as arthritis, autoimmune hemolytic anemia, scleroderma, and type 1 diabetes, are seen in X-linked agammaglobuline patients, in addition to chronic and/or recurrent infections (Azizi et al. 2016). Chronic granulomatous disease (CGD) is characterized by recurrent bacterial and fungal infections, as well as chronic inflammation of the gut and lungs. Neutrophils produce less nicotinamide adenine dinucleotide phosphate (NADPH) oxidase, which causes CGD (Arnold and Heimall 2017; Wolach et al. 2017).

Patients with midline granulomatous disease, an autoimmune condition usually seen in patients with granulomatosis with polyangiitis (Wegener's disease) and NK/T cell lymphomas, have hypomorphic RAG mutations. Finally, IgA deficiency has been linked to the development of autoimmune and atopic diseases in the past (De Ravin et al. 2010; Lee et al. 2012).

4.2 Autoimmunity Classification

Based on the main clinicopathologic features of each condition, autoimmune diseases may be classified into systemic, organ-specific, or localized autoimmune disorders (Fridkis-Hareli 2008).

1. SLE, Sjogren's syndrome, sarcoidosis, scleroderma, rheumatoid arthritis, and dermatomyositis are all examples of systemic autoimmune diseases. These conditions are often linked to autoantibodies to non-tissue specific antigens (Pasoto et al. 2019).
2. Localized syndromes affecting a single organ or tissue (Britannica 2021):

(a) Endocrinology: Type 1 diabetes, Hashimoto's thyroiditis, Addison's disease, Coeliac disease, Crohn's disease, and pernicious anemia are examples of gastrointestinal disorders (Van den Driessche et al. 2009; Krzewska and Ben-Skowronek 2016).
(b) Dermatologic: Vitiligo, Pemphigus vulgaris (Barker 2006).
(c) Haematology: Autoimmune haemolytic anemia, Idiopathic thrombocytopenic purpura (Giannotta et al. 2021).
(d) Neurology: Myasthenia gravis (Cetin et al. 2012).

4.3 Autoimmunity Factors

4.3.1 Genetic Variables

Some people have a hereditary tendency to autoimmune illnesses. This sensitivity is linked to several genes as well as other risk factors (Angum et al. 2020). Immunologists and geneticists have been working together to decode the immune system and discover the genetic and cellular foundations of its functioning and self-tolerance pathways, which can lead to autoimmunity when they are disrupted. Despite the central and peripheral tolerance systems' rigor and efficacy, a small percentage of potentially self-reactive lymphocytes escape into the periphery (Alriyami and Polychronakos 2021). Autoimmunity can be physiological, a transitory state with no clinical signs or pathology; yet their simple existence is insufficient to produce disease (Wang et al. 2015). The genetic component of the autoimmunity equation is well understood and established in today's genomic era, with the major risk loci having been found, in contrast to the environmental component, which remains unknown. However, current knowledge does not explain the autoimmunity discrepancies seen in monozygotic twins living in the same environment (at least for early-onset diseases like T1D) and in certain inbred animal models housed in a stable, constant, and controlled environment and, thus, identical in genome and environment. Furthermore, it does not account for autoimmunity's delayed stochastic penetrance. Discordant observations could be explained by stochastic events. The presence of somatic mutations in the growing autoimmune cells as a common pathogenetic mechanism for all forms of autoimmunity is one possible stochastic occurrence. Somatic mutations, which originate as self-reactive cells that proliferate following interaction with antigen, may interfere with one component of a set of self-tolerance checkpoints designed to stop their continued proliferation, according to this idea (Goodnow 2007). Proliferation opens the door to further somatic events, which could allow these lineages to circumvent several regulatory checkpoints, resulting in autoimmunity. The fundamental biological pathways of autoimmunity and cancer pathogenicity are similar, according to this somatic mutation theory (Goodnow 2007). Furthermore, it may be the missing piece of the autoimmunity puzzle, and recent genetic discoveries allow for testing of this idea, opening new research pathways in the field of autoimmunity and its tailored, personalized treatments. Even though DNA replication is a high-precision machine, errors do occur occasionally, resulting in mutations. This causes ongoing changes in the DNA, allowing for evolution and adaptability. Germinal/meiotic mutations and somatic mutations are two types of mutations. Somatic mutations, unlike germinal mutations, are genetic changes that are not passed down to offspring. They can happen at any moment during development when a mutation in a progenitor cell is passed down to all daughter cells and numerous tissue types. Furthermore, the earlier somatic mutations occur during embryogenesis, the greater the number of progeny cell clones that carry it (Behjati et al. 2014; Bae et al. 2018). When a somatic mutation occurs in a single post-mitotic cell in a fully grown organism, its influence is minimal, but when it occurs in cell types with proliferative properties, mutant clones can result (Griffiths et al. 2000). Somatic mutations, which affect tumor suppressor genes and oncogenes, are well-known and well-established contributors to carcinogenesis, resulting in aberrant morphologies and an escape from proliferative controls (Hanahan and Weinberg 2000; Watson et al. 2013). It also aided therapeutic progress by discovering pharmacological targets that are critical in illness causation (Hanahan and Weinberg 2000). Gleevec, for example, is a medication that acts as a specific inhibitor of the BCR-ABL tyrosine kinase, which is involved in chronic myeloid leukemia (CML) causing fusion transcript (Hernandez-Boluda and Cervantes 2002). The credibility of the somatic mutation theory in autoimmunity causation is determined by the prevalence of somatic mutations, particularly in autoimmune-relevant hematopoietic lineages (Alriyami et al. 2019). The study of minor genomic regions, like the PIG-A and HPRT genes (Araten et al. 2005), was the first step in attempting to quantify the rate of somatic mutations. Recent single-cell sequencing investigations in non-hematopoietic cells indicated that somatic mutations are prevalent in normal cells, with frequencies ranging from 3.5109 mutations/bp/division (in the small intestine) to 1.57107 mutations/bp/division (in the skin) (Blokzijl et al. 2016; Werner and Sottoriva 2018). Somatic mutations in non-hematopoietic cells are expected to rise steadily with age, at a rate of about 40 new mutations each year (Blokzijl et al. 2016). Somatic mutations in hematopoietic stem cells and progenitor cells can cause a mosaic condition in hematopoietic cells, which can be detected in the complete peripheral blood of healthy people (Young et al. 2016). Somatic mutations in genes with cancer implications, such as DNMT3A, TET2, and ASXL1, can arise in healthy people, predisposing them to hematological malignancies (Jaiswal et al. 2014).

4.3.2 Gender

Many complex features or diseases, such as infectious and autoimmune disorders, cancer, xenobiotic exposure, neurodevelopmental and neurodegenerative diseases, and vaccine outcomes, reveal a difference in vulnerability between men and women. In general, the female immune system responds to viruses more effectively. However, this might lead to over-reactive immune responses, which could explain why women are more likely to develop autoimmune disorders, as well as why females may have more negative vaccination side effects than males. Although sex hormone effects have been the focus of most research on gender differences in immunological responses, other mechanisms have been proposed, including cellular mosaicism, skewed X chromosome inactivation, genes escaping X chromosome inactivation, and miRNAs encoded on the X chromosome (Migliore et al. 2021).

Gender is determined by a variety of factors, including culture, ethnicity, social status, religion, and many more (Oertelt-Prigione and Mariman 2020), in addition to the gonosome makeup (normally determining the genetic sex, except for disorders of sexual development). As a result, gender, like age or environmental exposures, can be considered an environmental component (intrinsic) of susceptibility, influencing the risk of numerous complex diseases (Migliore et al. 2021).

4.3.3 Factors in the Environment

A rise in autoimmune disorders is posing a global socioeconomic problem. Despite the fact that predisposing genetic risk has been found, environmental factors have a substantial role in disease development and propagation. Changes in food habits are one of the most visible Western lifestyle characteristics probably connected with the rise in autoimmune illnesses, alongside greater hygiene and a significant reduction in infections. Growing data suggests that a typical "Western diet," which is high in saturated fat and salt, as well as linked diseases, can have a significant impact on local and systemic immune responses in physiologic and autoimmune illnesses like multiple sclerosis (Jörg et al. 2016).

Climate, stress, occupation, cigarette smoking, and food are only a few of the environmental factors that have been claimed to cause autoimmune illnesses like MS (Marrie 2004). It's worth noting that eating "Westernized cuisine," which includes a lot of salt, fat, protein, and sugar, has previously been linked to an increase in the prevalence of numerous diseases (Odegaard et al. 2012; Thorburn et al. 2014). The gut microbiota is strongly connected to the immune system and heavily implicated in immune regulation, and dietary changes have been extensively investigated, indicating a direct influence on immunological homeostasis and bacterial communities colonizing the gastrointestinal tract (GIT) (Conlon and Bird 2015). IBD has been linked to changes in the microbiome, including a decrease in variety and alterations. This has also been observed in other autoimmune illnesses that are not directly linked to the GIT. However, the pathways linking environmental variables to disease mechanisms, genetic predisposition, and the immune system are still poorly understood. Gaining a better understanding of the impact of the environment and microbiota on immunological homeostasis will be a valuable resource for better understanding the rising incidence of autoimmune diseases and developing novel therapeutic and preventative methods (Kuhn and Stappenbeck 2013; Salonen and de Vos 2014).

4.4 Autoimmunity Pathogenesis

The cause of most autoimmune disorders is unknown. An explanation for the pathogenesis of infection-related autoimmune disorders is offered. Antibodies are usually required when virulent pathogen infections produce large amounts of pathogen antigens. These antibodies form large antigen-antibody immunological complexes that some immunocompromised people are unable to remove. This results in inflammatory type III hypersensitivity symptoms, such as protease release, which destroys the basement membranes of the epithelium, mesothelium, and endothelium, expresses new immunogenic antigens from previously sequestered basement membrane constituents, and eventually induces new autoantibodies. If the first wave of protease attacks on basement membranes results in new autoantibodies, new uncleared antigen-antibody immune complexes, and type III hypersensitivity reactions, this can persist after the infection is over. Secreted proteases and other enzymes will have preferred substrates, and these proteases or other enzymes, alone or in combination with their processed protein substrates, can produce immunogenic antigens that elicit new autoantibodies and trigger autoimmune disorders. In summary, several autoimmune diseases can develop in immunocompromised people during long-term pathogen infections if they have two immune problems: (a) slow or weak initial immune responses that rely on antibodies, and (b) an inability to phagocytose the antigen-antibody immune complexes that result. These two immune issues, as well as the consequent immune system type III hypersensitivity reaction, can explain the causes of a variety of autoimmune diseases, including the most prevalent and rarest autoimmune disorders, as well as their similarities and distinctions (Roe 2021).

4.5 Molecular Mimicry

Right now, there are four significant measures that are contemplated to represent atomic mimicry (Peterson and Fujijami 2007; Tam et al. 2007): (1) "similitude between a host epitope and an epitope of a microorganism or natural specialist," (2) "recognition of antibodies or T-cells that cross-respond with the two epitopes in patients with AD," (3) "epidemiological connection between openness to the ecological specialist or organism and improvement of AD," and (4) "reproducibility of autoimmunity in a creature model after sharpening with the suitable epitopes either following contamination with the organism or openness to the natural specialist." Albeit these models have existed for quite a long while, they are trying to show in people for a few reasons. These incorporate the issues of inertness, the absence of enough epidemiologic force, the constraints of hereditary human investigations, the pertinence of innate murine models to outbred people, and the restricted innovation to separately consider the human white blood cell collection methodically. Also, there are different concerns regarding these models (Christen 2014). For instance, contamination might have happened a very long time before the beginning of illness, and not all tainted subjects foster an advertisement (Vatti et al. 2017). Moreover,

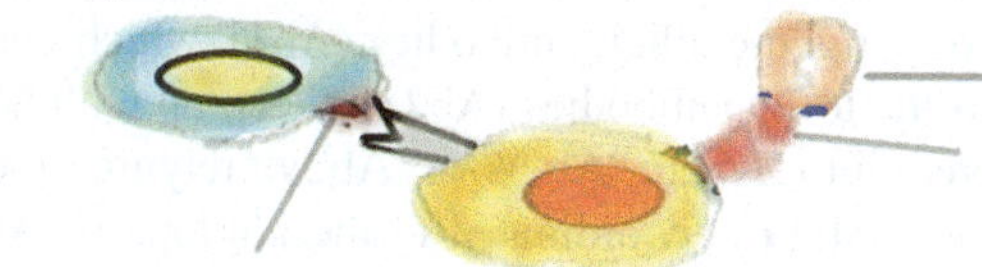

Fig. 4.1 Schematic diagram representing molecular mimicry in initiating autoimmune disease

people are tested with various contaminations across their lifetime, however, not this trigger autoimmunity, and some irresistible specialists might can possibly repeal the advancement of Advertisements (Christen 2014). Four kinds of sub-atomic mimicry have been proposed (Yuki 2007; Rodriguez et al. 2018), including: (1) Type 1: "complete character at the protein level between a microorganism and its host" (e.g., A human protein captured by the infection and introduced as an antigen), (2) Type 2: "homology at the protein level between a microorganism and its host, of a protein encoded by the microorganism," (3) Type III: "normal or comparable local or glycosylated amino corrosive successions or epitopes divided among the microorganisms or natural specialists and its host," and (4) Type 4: "primary similitudes between the organism or ecological specialists and its host." Albeit all the previously mentioned systems have been contemplated (Floreani et al. 2016), type III is the most widely recognized announced for advertisement since it is the least demanding to examine. Coming up next is a halfway posting of advertisements in which sub-atomic mimicry has been analyzed (Fig. 4.1).

4.6 Idiotype Cross-Reaction

The resistant framework is a perplexing organization of specific cells and organs that perceives and responds against unfamiliar microorganisms while staying inert to have tissues. This capacity to self-endure is known as immunological resistance. Immune system illness happens when the resistant framework neglects to separate among self and non-self-antigens and deliveries autoantibodies to assault our own cells. Hostile to idiotypic (against ID) antibodies are significant in keeping a fair idiotypic administrative organization by killing and restraining the emission of autoantibodies. As of late, hostile to ID antibodies have been progressed as an elective type of immunotherapy as they can explicitly target autoantibodies, cause less poisonousness and incidental effects, and could give dependable invulnerability (Pan et al. 2020). The idiotypic network was defined by Niels Jerne in 1974, including idiotypes (Ab1), against ID antibodies (additionally alluded to as Ab2) and antigen-introducing cells (Jerne 1974; Kieber-Emmons et al. 2012). As indicated by the organization hypothesis, he proposed that the resistant reaction may be managed by the reactions to idiotypes, which are special determinants of the immunoglobulin or white blood cell receptors (TCR) (Kieber-Emmons et al. 2012). These idiotypes (Ab1) can distinguish antigens and are perceived by Ab2 to keep up with homeostasis of the safe framework (Uner and Gavalchin 2006). During homeostasis, the presence of epitopes or antigenic determinants invigorate the creation of antigen-explicit antibodies (Ab1), which consequently actuate the creation of hostile to ID antibodies (Ab2) to keep up with harmony (Lemke 2005). Ab2 are ordered into Ab2α, Ab2β or Ab2γ, relying upon their antigen-restricting destinations. Ab2α is coordinated to the idiotope of Ab1, though Ab2γ is coordinated to the close to antigen epitope-restricting site idiotope of Ab1. Furthermore, Ab2β convey the inner picture restricting site to the paratope of Ab1 and tie to the complementarity deciding district of Ab1, showing that Ab2β mirrors basically towards Ab1. Ab2 can animate the creation of against hostile to ID (Ab3) which have comparable restricting limits as Ab1 (Shoenfeld 2004). This organization hypothesis permits the creation of against ID antibodies to control autoantibodies by killing and restraining the emission of autoantibodies, which ultimately assists with forestalling immune system sicknesses. Accordingly, numerous examinations zeroing in on Ab2 have been led to instigate target-explicit resistant reaction (Ladjemi 2012; Lule et al. 2017; Naveed et al. 2018). As of late, two new classifications of Ab2 were distinguished, known as Ab2δ and Ab2ε (Stanova et al. 2020). The idiotypic network has been demonstrated to play an essential part in the advancement of immune system illnesses, whereby lacking idiotypic guideline of autoantibodies has been viewed as a significant contributing component for autoimmunity. It is grounded that irregularities in the idiotypic organization might bring about the articulation and extension of autoantibodies. Because of immune system sicknesses, the idiotypic network is disturbed and favors the development of Ab1 autoantibodies that react to self-antigens (Uner and Gavalchin 2006). Autoantibodies, kept up with by both antigen and hostile to ID antibodies that tight spot to the antigen-restricting site of these autoantibodies, are normally found in low fixation in a solid grown-up (Pan et al. 2020). Thus, it is of general arrangement that the presence of raised coursing autoantibodies shows dysregulation of the humoral safe reaction. Since against ID antibodies can go about as the interior picture of antigen epitopes, they can seriously tie to the autoantibodies instead of the antigen. Thus, against ID antibodies help to kill and repress the creation of autoantibodies, coming about in a fair idiotypic network. Consequently, support of this harmony is fundamental to guarantee that the safe framework can viably battle against exogenous antigen without assaulting self-antigen that might actually prompt the improvement of immune system sicknesses (Pan et al. 2020) (Fig. 4.2).

4.7 Cytokine Dysregulation

Cytokines intercede and control insusceptible and fiery reactions. Complex cooperation exists between cytokines, irritation, and the versatile reactions in keeping up with homeostasis, wellbeing, and prosperity. Like the pressure reaction, the fiery response is essential for endurance and is intended to be custom fitted to the upgrade and time. An undeniable foundational provocative response brings about incitement of four significant projects: the intense stage response, the disorder condition, the aggravation program, and the pressure reaction, interceded by the

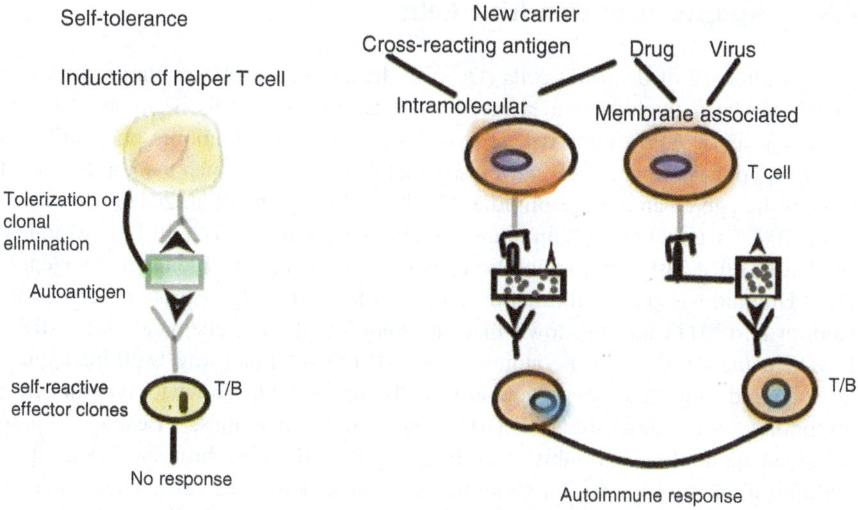

Fig. 4.2 Schematic diagram explaining the role of idiotype cross reaction in autoimmunity

hypothalamic-pituitary-adrenal hub and the thoughtful sensory system. Normal human illnesses like atopy/hypersensitivity, autoimmunity, ongoing diseases, and sepsis are described by a dysregulation of the favorable conditions versus calming and T aide (Th) 1 and Th2 cytokine balance. Ongoing proof additionally demonstrates the association of favorable to fiery cytokines in the pathogenesis of atherosclerosis and significant sorrow, and conditions like instinctive sort corpulence, metabolic disorder, and rest unsettling influences. During aggravation, the initiation of the pressure framework, through acceptance of a Th2 shift, shields the organic entity from fundamental "overshooting" with Th1/favorable to provocative cytokines. Under specific conditions, be that as it may, stress chemicals may really work with aggravation through acceptance of interleukin (IL)- 1, IL-6, IL-8, IL-18, growth corruption factor-alpha and C-receptive protein creation and through initiation of the corticotropin-delivering chemical/substance P-histamine hub. In this manner, a useless neuroendocrine-insusceptible interface related with irregularities of the "fundamental mitigating input" as well as "hyperactivity'" of the nearby supportive of fiery components might assume a part in the pathogenesis of atopic/unfavorably susceptible and immune system infections, weight, misery, and atherosclerosis. These irregularities and the disappointment of the versatile frameworks to determine aggravation influence the prosperity of the individual, including conduct boundaries, personal satisfaction, and rest, just as files of metabolic and cardiovascular wellbeing. These speculations require further examination, yet the appropriate responses ought to give basic experiences into systems fundamental an assortment of normal human insusceptible related illnesses (Elenkov et al. 2005).

4.8 Apoptosis of Dendritic Cells

The significance of dendritic cells (DCs) in the age of antiphospholipid syndrome (APS) has been uncovered in a few investigations. Apoptosis is involved in the pathogenesis of APS. Some debate exists, since uncovered anionic phospholipids during apoptosis might give B2GPI restricting destinations, which then, at that point prompt the age of antiphospholipid antibodies (aPL) (Tang et al. 2019). Bondanza et al. (2007) detailed that autoimmunity, including enemies of B2GPI IgG, would be created in mice just when apoptotic cells/B2GPI are infused alongside syngeneic DCs. Ubiquitin is one of the major pathways for intracellular protein corruption. Kool et al. (2011) tracked down that the ubiquitin-altering chemical, A20, stifles BMDC actuation through the atomic factor-κB (NF-κB) pathway. A20-inadequate DCs upgrade the take-up of apoptotic cells and antigen show to lymphocytes, prompting the downstream Th1 and Th17 reactions. Besides, these A20-inadequate DCs straightforwardly invigorate B cells, bringing about their multiplication and separation into immune response delivering cells. These creators likewise hereditarily erased A20 in mice and showed in vivo DC initiation and development and B and white blood cell actuation. Moreover, A20-lacking mice foster fundamental autoimmunity, including the creation of anticardiolipin IgG, thrombocytopenia, and fetal misfortune, all looking like APS. These discoveries copy the age of APS. One more investigation by Asano et al. (2004) showed that transformed milk fat globule-EGF-factor 8 (MFG-E8), which is initially communicated by DCs and macrophages to tie with phosphatidylserine (an anionic phospholipid) to advance phagocytosis, can restrain macrophages in the phagocytosis of apoptotic cells. At the point when intravenously infused, the freak protein actuates the creation of autoantibodies, including the anticardiolipin immunizer, which is upgraded by the concurrent infusion of syngeneic apoptotic thymocytes. Nonetheless, regardless of whether APS appearances created in these mice were not referenced. Kuwana et al. (2005) at first found that CD4+ immune system microorganisms, autoreactive to a mysterious peptide incorporating amino corrosive deposits 276–290 of B2GPI, are limited to HLA-DR53+ APS patients. They further treated blood monocyte determined DCs with B2GPI bound to anionic phospholipids. A coculture of these DCs and immune system microorganisms created autoreactive CD4+ white blood cells to a similar peptide in a HLA-DR-confined way ex vivo. In any case, we didn't recognize any examination showing the impact of DC disposal on the age of APS in mice. An immediate connection among DC and age of APS is subsequently deficient. Taken together, DC is fundamentally associated with the antigen shown by apoptotic cells, which could create autoreactive B and lymphocytes in APS (Tang et al. 2021).

4.9 Epitope Spreading or Epitope Drift

The broadening of the resistant reaction initiated by an antigen to new T-cell as well as neutralizer specificities over the span of an immune system sickness is known as "epitope spreading." This expanding of the invulnerable reaction can target epitopes either inside a similar antigen (intramolecular spreading) or another antigen (intermolecular spreading). Various variables are associated with the enlistment of epitope spreading, including the improved showcase of already obscure determinants under the nearby fiery/cytokine milieu, the arrival of self-antigens following tissue harm, the distinction in the size and energy of the epitope-explicit lymphocyte subsets and the job of B cells as antigen-introducing cells. Epitope spreading has commonly been conjured in the movement and chronicity of the underlying (intense) period of infection. In specific circumstances, notwithstanding, epitope spreading has been observed to be "defensive" or "illness controlling." Investigations of exploratory models have uncovered a "window" of helpful freedom even with sickness proliferating epitope spreading. Understanding the wonder of epitope spreading is significant for completely characterizing the pathogenesis of immune system sicknesses and for growing better immunotherapeutic methodologies for these issues (Shivaprasad et al. 2015).

4.10 Epitope Alteration or Cryptic Epitope Exposure

Insusceptible acknowledgment of self-proteins includes unmistakably in the early pathogenesis of immune system rheumatic illnesses like rheumatoid arthritis (RA), Sjoègren's syndrome (SS), systematic lupus erythematosus (SLE), and multiple sclerosis. The systems which furnish lymphocytes with admittance to such autoantigens are consequently key in making the opportunity for immune system reactions to create. It has for some time been imagined that the tissue or cell area of some self-proteins might establish that they are ordinarily "stowed away" from invulnerable acknowledgment, accordingly, decreasing their potential for autoantigenicity. As of late, this idea has been reached out to apply even to various epitopes inside a similar protein. Many investigations, enveloping a wide assortment of antigens, have shown that a few epitopes are not introduced for acknowledgment by T lymphocytes except if they are created in abnormally huge fixations or except if they are liberated from the arrangement of their local antigen. Epitopes for which this marvel happens are portrayed as enigmatic. There is expanding interest in the likelihood that crypticity might be a significant attribute of epitopes which are perceived by T lymphocytes in immune system pathogenesis (Warnock and Goodacre 1997).

References

Alriyami M, Polychronakos C (2021) Somatic mutations and autoimmunity. Cell 10:2056. https://doi.org/10.3390/cells10082056

Alriyami M, Marchand L, Li Q, Du X, Olivier M, Polychronakos C (2019) Clonal copy-number mosaicism in autoreactive T lymphocytes in diabetic NOD mice. Genome Res 29:1951–1961

Angum F, Khan T, Kaler J, Siddiqui L, Hussain A (2020) The prevalence of autoimmune disorders in women: a narrative review. Cureus 12(5):e8094. https://doi.org/10.7759/cureus.8094

Araten DJ, Golde DW, Zhang RH, Thaler HT, Gargiulo L, Notaro R, Luzzatto L (2005) A quantitative measurement of the human somatic mutation rate. Cancer Res 65:8111–8117

Arnold DE, Heimall JR (2017) A review of chronic granulomatous disease. Adv Ther 34:2543–2557. https://doi.org/10.1007/s12325-017-0636-2

Asano K, Miwa M, Miwa K, Hanayama R, Nagase H, Nagata S, Tanaka M (2004) Masking of phosphatidylserine inhibits apoptotic cell engulfment and induces autoantibody production in mice. J Exp Med 200:459–467

Azizi G, Ziaee V, Tavakol M, Alinia T, Yazdai R, Mohammadi H, Abolhassani H, Aghamohammadi A (2016) Approach to the management of autoimmunity in primary immunodeficiency. Scand J Immunol 85(1):13–29. https://doi.org/10.1111/sji.12506

Bae T, Tomasini L, Mariani J, Zhou B, Roychowdhury T, Franjic D, Pletikos M, Pattni R, Chen BJ, Venturini E et al (2018) Different mutational rates and mechanisms in human cells at pregastrulation and neurogenesis. Science 359:550–555

Barker JM (2006) Clinical review: type 1 diabetes-associated autoimmunity: natural history, genetic associations, and screening. J Clin Endocrinol Metab 91(4):1210–1217. https://doi.org/10.1210/jc.2005-1679

Behjati S, Huch M, van Boxtel R, Karthaus W, Wedge DC, Tamuri AU, Martincorena I, Petljak M, Alexandrov LB, Gundem G et al (2014) Genome sequencing of normal cells reveals developmental lineages and mutational processes. Nature 513:422–425

bio.libretexts.org (2021). https://bio.libretexts.org/Bookshelves/Microbiology/Book%3A_Microbiology_(Boundless)/12%3A_Immunology_Applications/12.5%3A_Immunity_Disorders%3A_Autoimmune_Diseases/12.5B%3A__The_Roles_of_Genetics_and_Gender_in_Autoimmune_Disease. Accessed 24 Mar 2021

Blokzijl F, de Ligt J, Jager M, Sasselli V, Roerink S, Sasaki N, Huch M, Boymans S, Kuijk E, Prins P et al (2016) Tissue-specific mutation accumulation in human adult stem cells during life. Nature 538:260–264

Bondanza A, Rovere-Querini P, Zimmermann VS, Balestrieri G, Tincani A, Sabbadini MG, Manfredi AA (2007) Requirement for dendritic cells in the establishment of anti-phospholipid antibodies. Autoimmunity 40:302–306

Britannica (2021). https://www.britannica.com/science/organ-specific-autoimmune-disease. Accessed 25 Mar 2021

Brodin P (2021) Immune determinants of COVID-19 disease presentation and severity. Nat Med 27:28–33. https://doi.org/10.1038/s41591-020-01202-8

Cetin H, Fulop G, Zach H, Auff E, Zimprich F (2012) Epidemiology of myasthenia gravis in Austria: rising prevalence in an ageing society. Wien Klin Wochenschr 124:763–768

Christen U (2014) Molecular mimicry. In: Shoenfeld Y, Meroni PL, Gershwin ME (eds) Autoantibodies, vol 2014, 3rd edn. Elsevier, San Diego, pp 35–42

Conlon MA, Bird AR (2015) The impact of diet and lifestyle on gut microbiota and human health. Nutrients 7(1):17–44

Cooper GS, Bynum ML, Somers EC (2009) Recent insights in the epidemiology of autoimmune diseases: Improved prevalence estimates and understanding of clustering of diseases. J Autoimmun 33:197–207

Cui Z, Zhao M-h, Segelmark M et al (2010) Natural autoantibodies to myeloperoxidase, proteinase 3, and the glomerular basement membrane are present in normal individuals. Kidney Int 78: 590–597

Davis R, Hollis T (2016) Autoimmunity: why the body attacks itself. Front Young Minds 4:23. https://doi.org/10.3389/frym.2016.00023

De Ravin SS, Cowen EW, Zarember KA, Whiting-Theobald NL, Kuhns DB, Sandler NG et al (2010) Hypomorphic rag mutations can cause destructive midline granulomatous disease. Blood 116(8):1263–1271. https://doi.org/10.1182/blood-2010-02-267583

Elenkov IJ, Iezzoni DG, Daly A, Harris AG, Chrousos GP (2005) Cytokine dysregulation, inflammation and well-being. Neuroimmunomodulation 12(5):255–269. PMID: 16166805. https://doi.org/10.1159/000087104

Floreani A, Leung PS, Gershwin ME (2016) Environmental basis of autoimmunity. Clin Rev Allergy Immunol 50:287–300. https://doi.org/10.1007/s12016-015-8493-8

Fridkis-Hareli M (2008) Hypothesis open access Immunogenetic mechanisms for the coexistence of organ-specific and systemic autoimmune diseases. J Autoimmune Dis 5:1. https://doi.org/10.1186/1740-2557-5-1

Giannotta JA, Fattizzo B, Cavallaro F, Barcellini W (2021) Infectious complications in autoimmune hemolytic anemia. J Clin Med 10:164. https://doi.org/10.3390/jcm10010164

Goodnow CC (2007) Multistep pathogenesis of autoimmune disease. Cell 130:25–35

Griffiths AJ, Miller JH, Suzuki DT, Lewontin RC, Gelbart WM (2000) Somatic versus germinal mutation. In: An introduction to genetic analysis. W. H. Freeman, New York

Gunjan K, Raza M, Tiwari BK (2018) Interferon-gamma (IFN-γ): exploring its implications in infectious diseases. Biomol Concepts 9:64–79

Halabi-Tawil M, Ruemmele FM, Fraitag S, Rieux-Laucat F, Neven B, Brousse N, De Prost Y, Fischer A, Goulet O, Bodemer C (2009) Cutaneous manifestations of immune dysregulation, polyendocrinopathy, enteropathy, X-linked (IPEX) syndrome. Br J Dermatol 160:645–651

Hanahan D, Weinberg RA (2000) The hallmarks of cancer. Cell 100:57–70

Hernandez-Boluda JC, Cervantes F (2002) Imatinib mesylate (Gleevec, Glivec): a new therapy for chronic myeloid leukemia and other malignancies. Drugs Today 38:601–613

Jaiswal S, Fontanillas P, Flannick J, Manning A, Grauman PV, Mar BG, Lindsley RC, Mermel CH, Burtt N, Chavez A et al (2014) Age-related clonal hematopoiesis associated with adverse outcomes. N Engl J Med 371:2488–2498

Jerne NK (1974) Towards a network theory of the immune system. Ann Immunol 125C:373–389

Jörg S, Grohme DA, Erzler M, Binsfeld M, Haghikia A, Müller DN, Linker RA, Kleinewietfeld M (2016) Environmental factors in autoimmune diseases and their role in multiple sclerosis. Cell Mol Life Sci: CMLS 73(24):4611–4622. https://doi.org/10.1007/s00018-016-2311-1

Kieber-Emmons T, Monzavi-Karbassi B, Pashov A, Saha S, Murali R, Kohler H (2012) The promise of the anti-idiotype concept. Front Oncol 2:1–12

Kool M, van Loo G, Waelput W, De Prijck S, Muskens F, Sze M, van Praet J, Branco-Madeira F, Janssens S, Reizis B et al (2011) The ubiquitin-editing protein A20 prevents dendritic cell activation, recognition of apoptotic cells, and systemic autoimmunity. Immunity 35:82–96

Krzewska A, Ben-Skowronek I (2016) Effect of associated autoimmune diseases on type 1 diabetes mellitus incidence and metabolic control in children and adolescents. Biomed Res Int 2016: 6219730. https://doi.org/10.1155/2016/6219730

Kuhn KA, Stappenbeck TS (2013) Peripheral education of the immune system by the colonic microbiota. Semin Immunol 25(5):364–369

Kusne Y, Christiansen M, Conley C, Gea-Banacloche J, Sen A (2021) Hemophagocytic lymphohistiocytosis secondary to disseminated histoplasmosis in rheumatologic disease. Case Rep Crit Care 2021:6612710. https://doi.org/10.1155/2021/6612710

Kuwana M, Matsuura E, Kobayashi K, Okazaki Y, Kaburaki J, Ikeda Y, Kawakami Y (2005) Binding of beta 2-glycoprotein I to anionic phospholipids facilitates processing and presentation of a cryptic epitope that activates pathogenic autoreactive T cells. Blood 105:1552–1557

Ladjemi MZ (2012) Anti-idiotypic antibodies as cancer vaccines: achievements and future improvements. Front Oncol 2:1–8

Lee YT, Chang YS, Lai CC, Chen WS, Yang AH, Tsai CY (2012) Natural killer (NK)/T-cell lymphoma mimicking granulomatosis with polyangiitis (Wegener's). Scand J Rheumatol 41(5): 407–408. https://doi.org/10.3109/03009742.2012.696684

Lemke H (2005) Idiotype network. In: Encyclopedic reference of immunotoxicology. Vohr H-W. Bayer HealthCare AG, Wuppertal, Germany, pp 307–311

Lule S, Colpak AI, Balci-Peynircioglu B, Gursoy-Ozdemir Y, Peker S, Kalyoncu U, Can A, Tekin N, Demiralp D, Dalkara T (2017) Behcet Disease serum is immunoreactive to neurofilament medium which share common epitopes to bacterial HSP-65, a putative trigger. J Autoimmun 84:87–96. https://doi.org/10.1016/j.jaut.2017.08.002

Marrie RA (2004) Environmental risk factors in multiple sclerosis aetiology. Lancet Neurol 3(12): 709–718

Marwaha AK, Leung NJ, McMurchy AN, Levings MK (2012) TH17 cells in autoimmunity and immunodeficiency: protective or pathogenic? Front Immunol 3:129. https://doi.org/10.3389/fimmu.2012.00129

Meffre E (2011) The establishment of early B cell tolerance in humans: lessons from primary immunodeficiency diseases. Ann N Y Acad Sci 1246:1–10

Migliore L, Nicolì V, Stoccoro A (2021) Gender specific differences in disease susceptibility: the role of epigenetics. Biomedicine 9:652. https://doi.org/10.3390/biomedicines9060652

Moulton VR, Tsokos GC (2015) T cell signalling abnormalities contribute to aberrant immune cell function and auto-immunity. J Clin Invest 125:2220–2227. [PubMed: 25961450]

Moulton VR, Suarez-Fueyo A, Meidan E, Li H, Mizui M, Tsokos GC (2017) Pathogenesis of human systemic lupus erythematosus: a cellular perspective. Trends Mol Med 23(7):615–635. https://doi.org/10.1016/j.molmed.2017.05.006

Mpakali A, Stratikos E (2021) The role of antigen processing and presentation in cancer and the efficacy of immune checkpoint inhibitor immunotherapy. Cancer 13:134. https://doi.org/10.3390/cancers13010134

Naveed A, Rahman SU, Arshad MI, Aslam B (2018) Recapitulation of the anti-idiotype antibodies as vaccine candidate. Transl Med Commun 3:1–7

Odegaard AO, Koh WP, Yuan J-M, Gross MD, Pereira MA (2012) Western-style fast food intake and cardiometabolic risk in an eastern country. Circulation 126(2):182–188

Oertelt-Prigione S, Mariman E (2020) The impact of sex differences on genomic research. Int J Biochem Cell Biol 124:105774

Pan SY, Chia YC, Yee HR, Fang Cheng AY, Anjum CE, Kenisi Y, Chan MK, Wong MB (2020) Immunomodulatory potential of anti-idiotypic antibodies for the treatment of autoimmune diseases. Fut Sci OA 7(2):FSO648. https://doi.org/10.2144/fsoa-2020-0142

Pasoto SG, de Oliveira A, Martins V, Bonfa E (2019) Sjögren's syndrome and systemic lupus erythematosus: links and risks. Open Access Rheumatol Res Rev 11:33–45. https://doi.org/10.2147/OARRR.S167783

Peterson LK, Fujijami RS (2007) Molecular mimicry. In: Shoenfeld ME, Gershwin ME (eds) Autoantibodies. Elsevier, Burlington

Rodriguez Y, Rojas M, Pacheco Y, Acosta-Ampudia Y, Ramirez-Santana C, Monsalve DM, Gershwin ME, Anaya JM (2018) Guillain-Barre syndrome, transverse myelitis and infectious diseases. Cell Mol Immunol 15:547–562. https://doi.org/10.1038/cmi.2017.142

Rodriguez-Rodriguez N (2015) Programmed cell death 1 and Helios distinguish TCR-alphabeta+ doublenegative (CD4-CD8-) T cells that derive from self-reactive CD8 T cells. J Immunol 194: 4207–4214. [PubMed: 25825451]

Roe K (2021) An explanation of the pathogenesis of several autoimmune diseases in immuno-compromised individuals. Immunology 93(3):e12994

Salonen A, de Vos WM (2014) Impact of diet on human intestinal microbiota and health. Annu Rev Food Sci Technol 5:239–262

Seetharam AB, Tiriveedhi V, Mohanakumar T (2010) Alloimmunity and autoimmunity in chronic rejection. Curr Opin Organ Transplant 15(4):531–536

Shivaprasad HV, Durai M, Moudgil K (2015) Epitope spreading in autoimmune diseases. Chapter 4. In: Infection and autoimmunity, pp 45–68

Shoenfeld Y (2004) The idiotypic network in autoimmunity: antibodies that bind antibodies that bind antibodies. Nat Med 10:17–18

Stanova AK, Ryabkova VA, Utekhin SV, Shoenfeld VJ, Churilov LP, Shoenfeld Y (2020) Anti-idiotypic agonistic antibodies: candidates for the role of universal remedy. Antibodies (Basel) 9(2):19

Tam CC, O'Brien SJ, Petersen I, Islam A, Hayward A, Rodrigues LC (2007) Guillain- Barre syndrome and preceding infection with campylobacter, influenza and Epstein-Barr virus in the general practice research database. PLoS One 2:e344. https://doi.org/10.1371/journal.pone.0000344

Tang KT, Hsieh TY, Chao YH, Li JP, Lan JL, Lin CC, Chen DY (2019) Apoptosis in patients with primary antiphospholipid antibody syndrome. Int J Rheum Dis 22:677–685

Tang K-T, Chen H-H, Chen T-T, Bracci NR, Lin C-C (2021) Dendritic cells and antiphospholipid syndrome: an updated systematic review. Life 11:801. https://doi.org/10.3390/life11080801

Thorburn AN, Macia L, Mackay CR (2014) Diet, metabolites, and "western-lifestyle" inflammatory diseases. Immunity 40(6):833–842

Uner A, Gavalchin J (2006) Idiotypes: encyclopedia of life sciences. Wiley, pp 1–5

Van den Driessche A, Eenkhoorn V, Van Gaal L, De Block C (2009) Type 1 diabetes and autoimmune polyglandular syndrome: a clinical review. Neth J Med Dec 67(11): 376–387. PMID: 20009114

Vatti A, Monsalve DM, Pacheco Y, Chang C, Anaya JM, Gershwin ME (2017) Original antigenic sin: a comprehensive review. J Autoimmun 83:12–21. https://doi.org/10.1016/j.jaut.2017.04.008

Velikova T, Georgiev T (2021) SARS-CoV-2 vaccines and autoimmune diseases amidst the COVID-19 crisis. Rheumatol Int 41:509–518. https://doi.org/10.1007/s00296-021-04792-9

Wang L, Wang FS, Gershwin ME (2015) Human autoimmune diseases: a comprehensive update. J Intern Med 278:369–395

Warnock MG, Goodacre JA (1997) Cryptic T-cell epitopes and their role in the pathogenesis of autoimmune diseases. Br J Rheumatol 36:1144–1150

Watson IR, Takahashi K, Futreal PA, Chin L (2013) Emerging patterns of somatic mutations in cancer. Nat Rev Genet 14:703–718

Werner B, Sottoriva A (2018) Variation of mutational burden in healthy human tissues suggests non-random strand segregation and allows measuring somatic mutation rates. PLoS Comput Biol 14:e1006233

Wolach B, Gavrieli R, de Boer M et al (2017) Chronic granulomatous disease: clinical, functional, molecular, and genetic studies. The Israeli experience with 84 patients. Am J Hematol 92(1): 28–36

Young AL, Challen GA, Birmann BM, Druley TE (2016) Clonal haematopoiesis harbouring AML-associated mutations is ubiquitous in healthy adults. Nat Commun 7:12484

Yuki N (2007) Ganglioside mimicry and peripheral nerve disease. Muscle Nerve 35:691–711. https://doi.org/10.1002/mus.20762

Zarrin AA, Bao K, Lupardus P et al (2021) Kinase inhibition in autoimmunity and inflammation. Nat Rev Drug Discov 20:39–63. https://doi.org/10.1038/s41573-020-0082-8

Autoimmunity and Diseases 5

5.1 Introduction

This chapter introduces several diseases that can be mediated by autoimmunity including diabetes, cancers, and others.

5.2 Diabetes Type 1 and Autoimmunity

T1D is caused by the immune system's destruction of insulin-producing beta cells, leading to a lifelong inability to maintain glucose homeostasis (Thomas et al. 2018). Insulin replacement therapy is currently the only treatment option. Although the basic characteristics of the disease are well accepted, several crucial uncertainties persist. While it has been known for many years that the rate of illness progression can differ significantly between individuals, the exact reason for this remains unknown. It was once thought that those with long-term T1D had few, if any, functioning beta cells left. However, it is now clear that this assumption was incorrect and that a considerable proportion of beta cells can survive persistent autoimmunity and produce detectable levels of insulin and C peptide decades after the start of clinical symptoms in many people (Keenan et al. 2010). Furthermore, histological examinations of the human pancreas at various stages of disease show that immunologic damage is not necessarily uniform (Atkinson et al. 2011). These findings show that individual beta cells may have cell-intrinsic variations in their responsiveness to a proinflammatory environment and/or vulnerability to autoimmune attack. Although there is significant evidence that aberrant beta cells increase during prediabetes in both animal models (Tersey et al. 2012) and people (Nyalwidhe et al. 2017), this remains largely a matter of hypothesis. However, it is unclear whether they are primarily a pathogenic shift that promotes autoimmunity (Sever et al. 2011) or a preventative reaction to prevent "collateral damage" during islet inflammation. Mitochondria play a key role in beta cell homeostasis, and there

A. J. Alkhatib, *The Role of Microbes in Autoimmune Diseases*,
https://doi.org/10.1007/978-981-19-1162-0_5

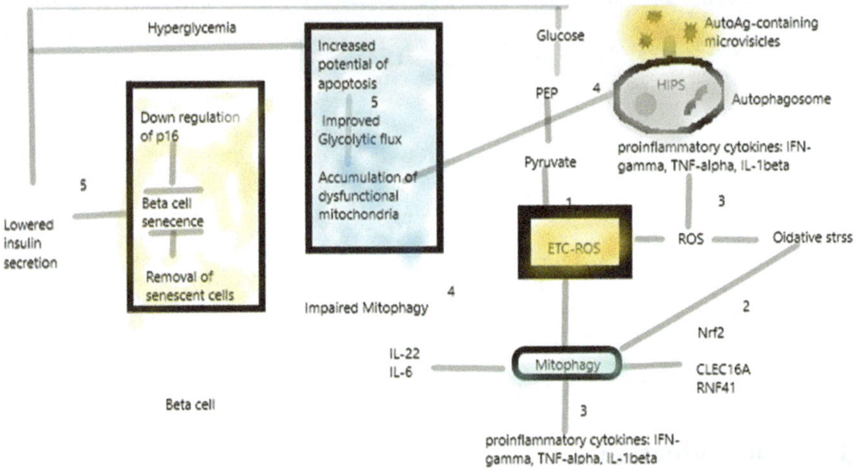

Fig. 5.1 Mitochondrial dysfunction could play a role in T1D beta cell failure. (1) Reactive oxygen species (ROS) are produced during oxidative phosphorylation, which can cause oxidative stress and damage. (2) Oxidative stress activates NRF and promotes protective responses, such as an increase in mitophagy. (3) Exposure to proinflammatory cytokines or other pathogenic stresses raises ROS levels, potentially overloading the detoxification system. (4) Impaired mitophagy will result in the accumulation of dysfunctional mitochondria and a global inhibition of flux through endo-lysosomal pathways, promoting the generation and/or secretion of immunogenic microvesicles containing autoantigens such as Hybrid Insulin Peptides (HIPs), forming a toxic positive feedback loop. (5) Beta cell survival adaptation mechanisms will become irreversible over time, resulting in beta cell malfunction, increased senescence, reduced insulin output, and eventually apoptosis. Known pathways are indicated by solid lines, whereas potential paths are indicated by dotted lines. Phosphoenolpyruvate (PEP) and Electron Transport Chain (ETC) are other abbreviations

is plenty of evidence that their malfunction is linked to type 2 diabetes (T2D) (Mulder and Ling 2009). Mitochondrial dysfunction has also been linked to T1D pathogenesis, primarily in the context of increased immune cell activity (Chen et al. 2017) and the production of reactive oxygen species (ROS) that can cause or exacerbate oxidative stress (Chen et al. 2018). Mitochondrial reactive oxygen species (ROS) is a required by-product of beta cells' central function of secreting insulin in response to glucose. Beta cells have multiple mechanisms to counteract the toxic effects of ROS, which are controlled by the transcription factor NF-E2-Related Factor 2 (NRF2; also known as nuclear factor, erythroid 2 like 2 (NFE2L2)) (Baumel-Alterzon et al. 2021), but mitochondrial ROS may still be a source of vulnerability that can worsen pathological conditions. The key role of mitochondria in beta cells is summarized in this study, as well as probable methods by which mitochondrial function and dysfunction in endocrine and immune cells may contribute to T1D pathology (Fig. 5.1).

Mitochondria are particularly vulnerable to damage from reactive oxygen species (ROS) and other forms of cellular stress (Ali et al. 2020), which can result in organelle malfunction and eventual loss of integrity. Mitophagy is a crucial defensive process that is significantly stimulated by oxidative stress and may eventually trigger apoptosis (Bock and Tait 2020). Because of this critical function, multiple lines of evidence relate mitophagy to inflammation and autoimmunity, which are implicated in the etiology of autoimmune illnesses such as T1D (Xu et al. 2020). The sensor kinase PINK1 (Bingol and Sheng 2016) triggers mitophagy by accumulating on the outer membrane of damaged mitochondria because of poor import of freshly produced molecules. Dimerization and autophosphorylation are enabled by reduced import, resulting in the fully active enzyme. Phosphorylation of serine-65 in the ubiquitin-like domain of the E3 ubiquitin ligase PARKIN and an equivalent residue in free ubiquitin by active PINK1 converts PARKIN to the active conformation, allowing attachment to mitochondria via phospho-ubiquitin modified surface proteins (Xiao et al. 2017). Active PARKIN then polyubiquitinates various mitochondrial outer membrane proteins, forming substrates for autophagy receptors like NDP52 and OPTN, which encourage the formation of phagophores around the injured organelle (Bingol and Sheng 2016). To finish the process, the closed autophagosome unites with lysosomes (Trivedi et al. 2020). Defective mitochondria can be recovered in addition to mitophagy by selective removal of damaged proteins/protein complexes in mitochondria-derived vesicles (MDVs) that bud from the outer membrane (Bingol and Sheng 2016). MDVs are likewise generated by PINK1 and PARKIN, but their distribution to late endosomes/lysosomes for destruction follows a different process than mitophagy and is not dependent on the core autophagic machinery. Because excessive mitophagy is harmful to beta cells, deubiquitinases such as USP15, USP30, and USP35, which can reverse PARKIN-mediated chain extension, limit autophagy receptor recruitment to mitochondria (Bingol and Sheng 2016).

5.3 Immune-Mediated Beta Cell Destruction and Autophagy

In rodent models of T1D, proinflammatory cytokines such as tumor necrosis factor-alpha (TNF-), interleukin-1 beta (IL-1), and interferon-gamma (IFN-gamma) have been identified as drivers of ROS generation and pathogenic ER stress (Eizirik et al. 2020). Although analogous processes are known to exist in human disease, the role of cytokines is less well understood. Furthermore, the exact involvement of autophagy in this process is still up for debate. Islets exposed to IFN- and IL-1 activated AMPK in response to ER stress in a rat model, resulting in decreased autophagic flux and altered lysosomal function, which contributed to beta cell death (Lambelet et al. 2018), leading to T1D in humans and mice (Muralidharan et al. 2021).

Genome Wide Association Studies (GWAS) have found several T1D susceptibility alleles that map to genes expressed in beta cells, leading to hypothesis that beta cells themselves contribute to T1D susceptibility and/or pathogenesis (Roep et al. 2021). We endeavored to describe the specific metabolic characteristics of the beta cell in this review, which allow it to efficiently secrete insulin in response to glucose stimulation and quickly adjust to changing dietary situations. Specialized mitochondrial functions, among these beta cell-specific processes, are particularly important in enhancing beta cell efficiency; yet they also create a milieu that is ripe for metabolic malfunction, and adaptive responses frequently become maladaptive with time. These findings back the theory that beta cells contribute to and/or perpetuate their own demise in the face of an autoimmune attack. To identify the extent to which mitochondrial dysfunction plays a role in the development of autoantigens that induce an autoimmune response, more research is needed (Kim et al. 2021).

Type 1 diabetes is now the most frequent type of diabetes in children, with estimates showing that 100,000 children are diagnosed each year (Patterson et al. 2019). Despite the availability of improved insulins, people with diabetes are still at a significant risk of catastrophic consequences, including cardiovascular death (Petrie and Sattar 2019). To improve the prognosis for the growing number of persons diagnosed with type 1 diabetes each year, new therapies are urgently needed (von Scholten et al. 2021).

The type 1 diabetes patient profile is changing and so is our understanding of the disease. The decrease of functional beta cell mass in the pancreatic islets of Langerhans is the main pathologic characteristic (Katsarou et al. 2017). According to some theories, the loss of functional beta cell mass occurs in a chain of events similar to a "assisted suicide" (Roep et al. 2021), in which the beta cell's death is caused by a combination of a dysfunctional beta cell that becomes more visible to the immune system, which then overreacts and destroys the beta cell.

Type 1 diabetes is frequently asymptomatic in its early stages (Stage 1); however, the onset of autoimmunity can often be detected in childhood, with circulating autoantibodies targeting insulin or other proteins such as GAD65, insulinoma-associated protein 2 (IA2), or zinc transporter 8 (ZNT8) (Katsarou et al. 2017). Asymptomatic dysglycemia (Stage 2) and, eventually, symptoms of hyperglycemia (Stage 3) result from insufficient or missing insulin secretion when a major amount of the beta cell mass has been dysfunctional or gone.

Type 1 diabetes is a polygenic condition, meaning that susceptibility loci and genetic variation all play a role in disease risk. The main susceptibility locus is the HLA region on chromosome 6, although numerous other loci across the genome have recently been linked to an increased risk of the disease (Pociot and Lernmark 2016). Non-genetic variables play a crucial role in causing or maintaining overt type 1 diabetes, according to research on monozygotic twins (Redondo et al. 2008), for whom the development of type 1 diabetes might differ significantly. Numerous attempts have failed to find such components, implying that no single pathogen is to blame. Enteroviruses and human herpesvirus-6 have been postulated as viral infections (Rodriguez-Calvo 2018; Sabouri et al. 2020). However, according to the "hygiene theory" (von Mutius 2007), research (mostly in animals) has revealed that various viral infections may inhibit the development of type 1 diabetes (Filippi and von Herrath 2005; Christen and von Herrath 2011). Exogenous insulins continue to be the cornerstone therapy option for those with type 1 diabetes (Beck et al. 2019). Novel and adaptable formulations, analogues, and delivery vehicles have been introduced since the discovery of insulin in 1921 (Beck et al. 2019).

5.4 Therapies That Target the Immune System

In type 1 diabetes, the overall goal of immune-focused therapy is to prevent or delay the loss of functional beta cell mass. The classic view of autoimmunity in type 1 diabetes has focused on systemic immune dysregulation and autoreactive T cells that have moved to the periphery and destroyed islets after eluding thymic selection. T cell-mediated "homicide" has been coined to describe this theory of type 1 diabetes development (Atkinson et al. 2011). As a result, current attempts have focused on cell or cytokine-directed therapies, which have proven to be effective in other autoimmune illnesses.

Autoantibodies to beta cell antigens, such as GAD65 and insulin, have prompted attempts to target B cell-related components. These efforts have had some success in animal models (Xiu et al. 2008), as well as in clinical trials, most notably with the anti-CD20 antibody rituximab, which depletes B cells. The effect of rituximab on beta cell function was detectable (Pescovitz et al. 2009), but it was temporary (Pescovitz et al. 2014), demonstrating that B cell-directed treatment alone does not appear to prevent or ameliorate beta cell autoimmune in a long-term manner. However, B cell-directed medicines have yet to be evaluated in the early stages of the disease, precluding judgments about their efficacy in delaying or even stopping development to later stages.

Tissue-resident memory T effector cells, which are thought to play a role in a variety of organ-specific autoimmune disorders such type 1 diabetes, are notoriously tough to eradicate. Rigby et al. (2015) evaluated alefacept, a T cell-depleting fusion protein that targets CD2 and, hence, memory T effector cells, in adolescents and young people with Stage 3 type 1 diabetes. Even though the trial did not reach its target enrollment, it reported a trend of advantages in terms of beta cell preservation, lower insulin requirements, and a low risk of hypoglycemia that lasted for the whole 15-month follow-up period.

Another proinflammatory cytokine that has been successfully targeted in autoimmune disorders is IL-6 (Kang et al. 2019). Although its role in type 1 diabetes is unknown, IL6 has been proposed as a potential target (Hundhausen et al. 2016). IL-6 has been found to protect beta cells from oxidative stress and is expressed constitutively by pancreatic alpha and beta cells, implying crucial physiological functions (Rajendran et al. 2020). Tocilizumab, a monoclonal antibody against the IL-6 receptor, was recently completed in the EXTEND Phase II trial in type 1 diabetes (ClinicalTrials.gov registration no. NCT02293837).

5.5 Diabetes Type 2 and Autoimmunity

The insulin signaling pathway can be inhibited by inflammation. Insulin resistance develops because of this, and type 2 diabetes develops as a result (T2DM). Obesity and insulin resistance are linked to a chronic yet undiagnosed inflammatory process that inhibits insulin action in most tissues and may impair pancreatic-cell function. The presence of monocytic cells and the chemokine and cytokine profiles caused by inflammation point to an innate immune response. However, new evidence suggests that adaptive immune system components may possibly be implicated. Because antigen specificity is required for adaptive immune response activation, some researchers believe T2DM develops from an innate immune response to an autoimmune illness (Velloso et al. 2013).

T2DM (type 2 diabetes mellitus) is one of the most common diseases in modern society, affecting more than 340 million people worldwide (Danaei et al. 2011). Obesity is by far the most important predisposing factor for T2DM, and the rapid rise in obesity prevalence is likely to parallel the rise in T2DM prevalence. The key environmental factor contributing to obesity is the consumption of energy-dense and fat-rich foods (Schulz et al. 2002). Over the last 20 years, clinical and molecular research have unequivocally shown the significance of diet-induced subclinical inflammation as an important link between obesity and T2DM (Shoelson et al. 2006). Toll-like receptors (TLRs) 2 and 4 are activated by dietary lipids, which cause endoplasmic reticulum stress (ERS) in adipose tissue, the liver, and the hypothalamus (Milanski et al. 2009; Pal et al. 2012). Inflammatory activity is induced by TLR and ERS signaling, which activates intracellular serine–threonine kinases that block insulin signaling (Cnop et al. 2012). TLR and ERS signaling also increase inflammatory gene transcription, culminating in the synthesis and secretion of cytokines including tumor necrosis factor (TNF) and IL-1, which boost intracellular inflammatory signaling and insulin resistance via an extracellular feed-forward loop (Gregor and Hotamisligil 2011). Saturated fats induce ERS, which leads to the activation of inflammatory mediators (Eizirik et al. 2013) and the malfunction and death of pancreatic cells in T2DM patients (Cnop et al. 2012; Eizirik et al. 2013).

Macrophages and TNF play a role in both inducing and maintaining the inflammatory activity connected to obesity, leading to the theory that insulin resistance is caused by an abnormal activation of the innate immune system (Osborn and Olefsky 2012). Other research, on the other hand, suggests that adaptive immune response components have a role in the advancement of the inflammatory response linked to obesity (Feuerer et al. 2009; Guarda et al. 2009). The link between the major histocompatibility complex (MHC) and the T-cell receptor (TCR) is required for the activation of an adaptive immune response, which needs the processing and presentation of specific antigens. Inflammation is not known to be triggered by infectious pathogens in either obesity or T2DM. As a result, endogenous antigens or exogenous substances (e.g., dietary components) may be presented, behave as antigens, or abnormally alter the immune response, activating subsets of lymphocytes that target host cells and tissues (Velloso et al. 2013). Obesity (and possibly T2DM) could have an autoimmune component if evidence of such a mechanism could be found. It is less clear whether T2DM has an autoimmune component in its etiology. T cells have yet to be identified as a potential autoantigenic target in this situation, but multiple studies have found potential targets for IgG antibodies linked to insulin resistance (Winer et al. 2011) and autoantibodies against pancreatic islet antigens in T2DM patients (Brooks-Worrell et al. 1999). Although glucose intolerance and insulin resistance develop in recipients of IgG antibody transfers in rodents, the insulin resistance phenotype is not observed to be conveyed during bone marrow transplantation (Winer et al. 2011).

T-lymphocyte suppression has not been shown to have a therapeutic or preventive effect in T2DM in people; nevertheless, inhibition of either B cells (Winer et al. 2011) or T lymphocytes (Winer et al. 2009) can slow the onset of obesity-related insulin resistance in mouse models. Although there is no strong evidence of autoimmune in animal models of T2DM, targeting adaptive immune system components such as IFN-expressing type 1 T helper cells and B lymphocytes can ameliorate insulin resistance (Winer et al. 2009, 2011). Notably, there is relatively little overlap between T1DM and T2DM candidate genes; only GLIS3 (which encodes a zinc finger protein) is linked to both disorders out of over 50 discovered (Dupuis et al. 2010).

Furthermore, T2DM is not linked to any of the recognized candidate genes for other inflammatory or autoimmune illnesses, such as IL23R, IL2RA, PTPN2, or several HLA alleles (Winkler et al. 2012). Most genes associated to obesity are thought to function on energy-balancing neuroendocrine circuits and have not been connected to innate or adaptive immunity (Speliotes et al. 2010). While these genetic polymorphisms account for just a small portion of the heritability of obesity and T2DM, they do not support the idea that autoimmunity plays a role in these two metabolic diseases. Low levels of cytokines and minor immune cell infiltration in metabolic tissues imply continuous but low-level inflammation, both of which are key elements of the inflammatory process in metabolic tissues. In contrast to the conventional inflammatory process seen in most autoimmune diseases, metabolic inflammation is not associated with the typical indications of inflammatory activity seen in people with obesity and/or T2DM, such as heat, oedema, redness, and discomfort (Calay and Hotamisligil 2013). Furthermore, metabolic inflammation causes a reduction in whole-body energy expenditure rather than an increase in it (Calay and Hotamisligil 2013). Despite these significant variations from traditional inflammatory processes, cryptic antigens may still be exposed because of prolonged low-level inflammation, leading to an adaptive autoimmune response, as previously postulated (Brooks-Worrell et al. 2012).

5.6 T2DM, B Cells, and Inflammation

Regardless of whether they develop T2DM, the inflammatory process in adipose tissue appears to be comparable in all insulin resistant patients with obesity (Barbarroja et al. 2012). Obese people only get diabetes mellitus when their pancreas cells can't compensate for their insulin resistance. The evidence for innate and adaptive immunity playing a role in cell failure in T2DM patients is mixed. The number of macrophages in pancreatic islets (measured by counting the number of CD68+ cells per islet) increased from 0.5 cells in non-diabetic people to 1.5 cells in T2DM patients (Richardson et al. 2009). The proportion of islets containing at least five immune cells increased from 0.6% in nondiabetic control individuals to 5.6% in patients with T2DM (Richardson et al. 2009). The presence of these macrophages has been linked to viral infection of islets (Richardson et al. 2009) and expression of human islet amyloid polypeptide (Masters et al. 2010); however, their role remains unclear. Microarray analysis of laser-capture microdissected β cells showed a minor increase (twofold to threefold) in levels of the chemokines CCL2, CCL11, CCL13, and CXCL1 and in the cytokines IL-1β and IL-8 (but not IFN-γ) in cells from patients with T2DM, compared with those from nondiabetic controls (Igoillo-Esteve et al. 2010). In patients with T2DM, the fraction of islets harboring at least five immune cells increased from 0.6% in nondiabetic controls to 5.6% (Richardson et al. 2009). The existence of these macrophages has been associated to viral infection of islets (Richardson et al. 2009) and expression of human islet amyloid polypeptide (Masters et al. 2010), but their function is unknown. In cells from patients with T2DM, microarray analysis of laser-capture microdissected cells revealed a slight rise (twofold to threefold) in levels of the chemokines CCL2, CCL11, CCL13, and CXCL1, as well as the cytokines IL-1 and IL-8 (but not IFN-) compared to nondiabetic controls (Igoillo-Esteve et al. 2010).

Mild inflammation is indicated by these cytokine and chemokine profiles, which is consistent with an innate but not an adaptive immune response. In vitro, exposing human islets to the saturated fatty acid palmitate or synthetic ERS inducers induced a comparable response, which was dependent on IL-1 production (Igoillo-Esteve et al. 2010). Palmitate-induced inflammation was reduced by blocking IL-1 using an IL-1 receptor antagonist, but not cell apoptosis (Igoillo-Esteve et al. 2010). Microarray analysis of islet cells from T2DM patients connected a collection of coexpressed genes (many of which were IL-1-related genes) to lower insulin secretion, which backed up these findings (Mahdi et al. 2012). Adults with latent autoimmune diabetes mellitus, a disease that shares many characteristics with T1DM and is sometimes misdiagnosed as T2DM, have an autoimmune reaction to pancreatic cells (Tuomi et al. 1993). Responses to islet-reactive T cells were measured in certain autoantibody-negative patients with T2DM by evaluating mononuclear cell responses to human islet proteins blotted onto nitrocellulose (Brooks-Worrell and Palmer 2012). Due to a lack of control tissues and particular identification of target antigens, this method of measuring T-cell activation is controversial, as it may represent unspecific inflammation rather than autoimmunity. Other research groups utilizing traditional tests based on particular cell antigens need to corroborate the data demonstrating T-cell responses against islet proteins in T2DM patients. Unlike T1DM, where nearly all cells are finally killed by the autoimmune response, T2DM patients have a small drop in cell mass (approximately 30–60%). This significant observation implies that the processes of cell loss in T1DM and T2DM are distinct (Cnop et al. 2005).

5.7 The Involvement of Autoimmunity in Cancer

A relationship between chronic autoimmune gastritis and stomach cancer has been observed in many research, and changes in the immune response of people with autoimmune disorders may predispose to malignancies. Precursor lesions include intestinal metaplasia with dysplasia of the gastric corpus-fundus mucosa and hyperplasia of chromaffin cells, which are common in late-stage autoimmune gastritis. The development of two forms of gastric neoplasms has been linked to autoimmune gastritis: intestinal type and type I gastric carcinoid. We look at the link between autoimmune gastritis and gastric cancer, as well as other autoimmune characteristics found in gastric neoplasms (Bizzaro et al. 2018).

Immune dysregulation is thought to have a pathogenic role in the development of autoimmunity and neoplasia, and autoimmune disorders have been reported in patients with cancer. Antinuclear antibodies, a hallmark of many autoimmune rheumatic diseases, have been found in the sera of patients with malignant tumors (Solans-Laqué et al. 2004); anti-La antibodies, which are commonly found in the sera of patients with Sjögren's syndrome, and anti-CENP-B antibodies, a marker of systemic sclerosis, have also been found in the sera of patients with breast cancer (Toubi and Shoenfeld 2007). Anti-dsDNA antibodies, which are diagnostic and prognostic in systemic lupus erythematosus (SLE), have also been found in the sera of patients with various types of cancer (Lv et al. 2005), and the presence of rheumatoid factor has been linked to a poor prognosis in various types of cancer, including gastrointestinal cancer (Konstadoulakis et al. 1994). Organ-specific antibodies, such as anti-smooth muscle antibodies, anti-parietal cell antibodies, and anti-thyroid antibodies, have also been reported in malignancies (Molander et al. 2000).

Patients with autoimmune illnesses, on the other hand, have a higher risk of developing cancer (Tomer and Shoenfeld 2000). According to the Bradford-Hill (1965) postulates, rheumatoid arthritis, SLE, Sjögren's syndrome, and celiac disease have been linked to lymphoproliferative diseases (Mellemkjaer et al. 1997), idiopathic inflammatory myositis with solid tumors (Villa et al. 2000), and systemic sclerosis with malignant neoplasms (Moinzadeh et al. 2014). Furthermore, recent research has revealed that autoimmune gastritis can progress to neoplastic transformation in as little as 10% of cases, and that autoimmune gastritis can be considered a pre-neoplastic condition with a 0.3% yearly risk of gastric cancer (Toh 2014). Autoimmune Gastritis (AIG) is an organ-specific disease characterized by a chronic inflammation of the stomach mucosa that progresses to atrophic gastritis, resulting in malabsorption of essential nutrients and, eventually, microcytic iron-deficient anemia (Marignani et al. 1999) or pernicious anemia due to vitamin B12 deficiency (Bizzaro and Antico 2014). The parietal and main cells of the mucosa may be replaced by mucus-producing cells, similar to those seen in the intestine, as the lesion advances. Intestinal metaplasia and spasmolytic polypeptide-expressing metaplasia (SPEM) are two kinds of metaplasia linked to gastric carcinogenesis in humans. Intestinal metaplasia goblet cells express suitable intestinal markers such as Muc2 and Trefoil factor 3 (TFF3), but mucous metaplastic lineages in SPEM have morphological traits more akin to deep antral gland cells or Brunner's glands, with Muc6 and Trefoil factor 2 expression (TFF2). Recent research supports the formation of SPEM because of transdifferentiation from mature main cells following the loss of parietal cells (Weis and Goldenring 2009). Both intestinal metaplasia and SPEM have been linked to the development of gastric cancer of the intestinal type

(Kokkola et al. 1998). AIG, like other autoimmune diseases, is more frequent in women than in men (3:1 ratio). AIG is usually asymptomatic until the mucosa has advanced to the point of atrophy and/or dysplasia (Dixon et al. 1996). As a result, AIG is an illness that is frequently underdiagnosed, with an estimated prevalence ranging from roughly 2% in the third decade to 12% in the eighth (Weck and Brenner 2006). Patients with various autoimmune disorders, particularly autoimmune thyroid diseases (AITD) and type 1 diabetes (T1DM), have a higher prevalence (Van den Driessche et al. 2009). Multiple autoimmune disorders (MAS) types 3B and 4 are defined by these connections (Betterle and Presotto 2008).

Chronic autoimmune gastritis (type A) differs from type B gastritis caused by Helicobacter pylori (H. pylori) infection in terms of etiology and histology (Strickland and Mackay 1973). Because inflammatory aggressiveness targets the cells of the oxytocin glands, AIG is confined to the gastric body and fundus, unlike *H. pylori* gastritis, which is mostly found in the antrum (Toh et al. 2000). However, during *H. pylori* infection, a specific form of AIG may emerge in genetically predisposed people (Weck and Brenner 2008). Anti-parietal cell antibodies were found in 20–30% of patients with *H. pylori* infection, and anti-*H. pylori* antibodies were found in patients with AIG, suggesting that *H. pylori* and gastric autoimmunity are linked (Faller and Kirchner 2005).

H. pylori infection may cause AIG by molecular mimicry and/or epitope spreading mechanisms; a high degree of similarity has been found between the Hp urease component and the subunit of gastric ATPase (Amedei et al. 2003). The activation of gastric Th1 cells in response to *H. pylori* wall peptides that cross-react with gastric H +K+-ATPase triggers an inflammatory response in which T-cell-derived IFN allows parietal cells to act as APCs and become targets of cross-reactive epitope recognition, resulting in death or apoptosis. Apoptotic parietal cells would thus allow T lymphocytes specific to private gastric ATPase epitopes to cross-prime (Plebani and Basso 2009). Although histological healing of the body mucosa has been described in patients who have had *H. pylori* removed (Ohkusa et al. 2001), a direct link between *H. pylori* infection and AIG is still debated (De Block et al. 2002). To this purpose, it should be emphasized that while the bacteria is present in the early stages of gastritis, it is no longer detectable in the atrophic stage due to hypocloridry and mucosal degradation, which creates an environment unfavorable for *H. pylori* survival.

5.8 Gastric Cancer and Autoimmune Gastritis

Patients with AIG have a higher rate of stomach neoplasms than the overall population (Vannella et al. 2013). According to prospective studies, 4–9% of individuals with AIG, or its more severe counterpart pernicious anemia, develop gastric carcinoid tumors, which are 13 times more common than in control people (De Block et al. 2008). Furthermore, AIG development to atrophic gastritis, which is associated with intestinal metaplasia, may predispose more than 10% of individuals to stomach cancer (De Block et al. 2008). Individuals with AIG/pernicious anemia had a threefold increased risk of developing not only stomach carcinoid and adenocarcinomas, but also small intestinal adenocarcinomas and esophageal carcinoid and adenocarcinomas, according to two recent studies, one involving over 4.5 million adult male veterans admitted to US Veterans Affairs hospitals in the United States (Landgren et al. 2011) and the other involving nine million individuals from Sweden (Hemminki et al. 2012).

Taken together, there is evidence that patients with autoimmune gastritis have a greater rate of stomach neoplasms than the normal population. Chronic inflammation stimulates gastric cells to produce inflammatory cytokines, which play a role in regulating oxyntic atrophy, hyperplasia, metaplasia, and progression to gastric cancer by up-regulating expression of progenitor cells, according to numerous studies in humans and mouse models of gastritis. Recent findings on the role of stem cells in gastric cancer may provide light on the molecular mechanism of carcinogenesis, paving the way for the development of new therapeutic approaches to treat early-stage gastric cancer (Bizzaro et al. 2018).

Autoimmune cytopenias (AICy) and autoimmune disorders (AID) are common complications of lymphoid and myeloid neoplasms, and they can be difficult to diagnose and treat. While AIHA and ITP are well-known, other AICy (autoimmune neutropenia, aplastic anemia, and pure red cell aplasia) and AID (systemic lupus erythematosus, rheumatoid arthritis, vasculitis, thyroiditis, and others) are less well-known. This review examines the literature over the past 30 years on the prevalence of AICy/AID in various onco-hematologic diseases. CLL, lymphomas, multiple myeloma, myelodysplastic syndromes (MDS), chronic myelomonocytic leukemia (CMML), myeloproliferative neoplasms, and acute leukemias are among the latter. AICy are found in up to 10% of CLL cases and at higher rates in some subtypes of non-Hodgkin lymphoma, while they are seen in less than 1% of low-risk MDS and CMML cases. Immune-mediated hemostatic diseases (acquired hemophilia, thrombotic thrombocytopenic purpura, and anti-phospholipid syndrome) been described in up to 30% of myeloid and lymphoid individuals and can be severe and deadly. AICy/AID are also seen in roughly 10% of patients following hematopoietic stem cell transplants or novel checkpoint inhibitor medication. Aside from the diagnostic challenges, this AICy/AID may complicate the clinical management of patients who are already immunocompromised (Barcellini et al. 2021).

Autoimmune consequences in hematologic malignancies are becoming more well-known. Autoimmune cytopenias (AICy) in the periphery, such as autoimmune hemolytic anemia (AIHA) and immune thrombocytopenia (ITP), are well-known lymphoproliferative disorders (LPD) consequences (Barcellini et al. 2020a, b; Fattizzo and Barcellini 2020). Other autoimmune diseases (AIDs) that have received less attention include organ-specific disorders affecting the endocrine glands, liver, and gut, as well as systemic AIDs (i.e., systemic lupus erythematosus, SLE, rheumatoid arthritis, RA, and antiphospholipid syndrome, APS). Furthermore, there is evidence of several autoimmune events, such as autoantibodies against various autologous proteins, which may or may not signal the onset of disease. Due to overlapping diseases such as chemotherapy, bone marrow invasion, and transfusion support, diagnosing AICy in hematologic malignancies can be difficult. The direct antiglobulin test (DAT) or Coombs's test for the diagnosis of AIHA is known to be negative in up to 10% of cases, while anti-platelet and anti-neutrophil antibody detection has low sensitivity (Barcellini et al. 2020a, b; Fattizzo and Barcellini 2020).

Autoimmunity is caused by a complex interaction of genetic and environmental variables, such as infections and medications. Primary immunodeficiencies are known to have a high prevalence of autoimmune illnesses as well as an abnormal immune response to infections (Rizvi et al. 2020). In addition, autoimmune patients are more susceptible to infections, probably because of immunosuppressive medication, setting off a vicious cycle that exacerbates immunological dysregulation. Drugs are a well-known cause of AID, with procainamide and hydralazine for SLE and methyldopa for AIHA being reported in the past. New biological modulators, such as tumor necrosis factor (TNF) inhibitors, other cytokine inhibitors, and immune checkpoint inhibitors, have lately been linked to autoimmune problems. The higher risk of AID/AICy associated with solid organ and hematopoietic stem cell transplants (HSCT) stems from the severe immunological storm caused by the transplant as well as the therapy-induced immunodepression and subsequent infections (Barcellini and Fattizzo 2020; Barcellini et al. 2020a, b).

5.9 Autoimmunity and Neurological Diseases

Several previously poorly understood neurological illnesses have had their autoimmune origin explained by remarkable findings over the last two decades. Autoimmune nervous system illnesses can affect any portion of the nervous system, including the brain and spinal cord (central nervous system, CNS), as well as peripheral nerves, the neuromuscular junction, and skeletal muscle (peripheral nervous system, PNS). This comprehensive introduction of this rapidly growing topic discusses the elements that can cause self tolerance to break down and the onset of autoimmune disease in some people. The pathophysiological underpinnings and clinical aspects of autoimmune nerve system illnesses are next discussed, with a focus on the features that are critical to recognize for proper clinical diagnosis (Bhagavati 2021).

Many new autoimmune illnesses of the neurological system have been found and their pathophysiology has been revealed thanks to significant discoveries made during the previous two decades. Figures 5.2 and 5.3 describe the clinical and pathophysiological features of CNS autoimmune illnesses, while Fig. 5.4 describes the clinical and pathophysiological features of PNS autoimmune diseases.

Autoantibodies directed against intracellular neuronal antigens (typical paraneoplastic diseases) are one type of autoimmune condition that affects the nervous system (Mckeon and Pittock 2011). Antibodies to certain proteins (such as anti-Hu, anti-Ri, and anti-Ma) are virtually always linked to the existence of a tumor. These illnesses are thought to be induced by an immune response that is misdirected against tumor proteins (onconeural antigens) that are also produced by neurons. Autoantibodies that are abnormally directed against intracellular neuronal

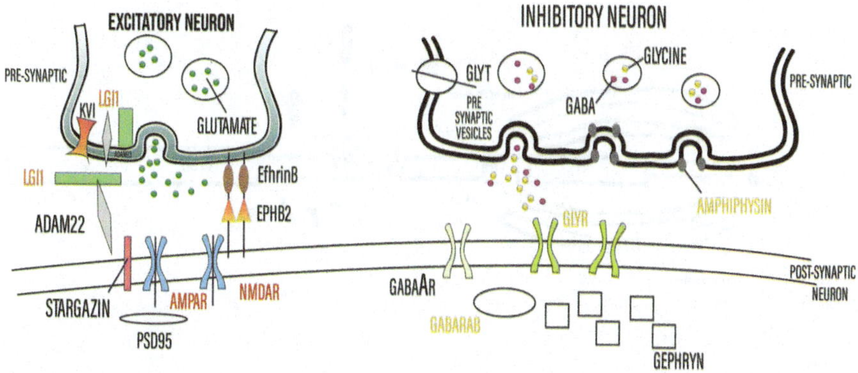

Fig. 5.2 Autoimmune disorders of the central nervous system. Synapses that are excitatory: Anti-NMDA receptor (NMDAR) antibodies bind and cross-link GluN1 subunits, disrupting NMDAR's interaction with ephrin type B receptor 2 (EPHB2), resulting in NMDAR internalization and decreased glutamergic transmission. Autoantibodies to leucine rich glioma inactivated protein 1 (LGI1), a neuronal glycoprotein released into the synapse, induce LGI1 encephalitis. LGI1 controls AMPA receptor trafficking and presynaptic Kv1, voltage-gated potassium channel activity via interacting with presynaptic disintegrin and metalloproteinase domain-containing protein 2 (ADAM23) and postsynaptic ADAM22. PSD95 is an acronym for post-synaptic density protein 95. Other receptors to be targeted include mGluR1/5; AMPAR Autoantibodies to the GABA type A receptor (GABAAR) on inhibitory synapses can cause encephalitis and intractable seizures. Autoantibodies to the glycine receptor (GlyR) induce painful spasms, stiffness, and myoclonus in the brain (PERM). Autoantibodies to amphiphysin, a presynaptic protein present in all synapses and critical in clathrin-mediated endocytosis, can produce stiff person syndrome by reducing the quantity of presynaptic vesicles packed with neurotransmitter accessible for exocytosis. Inhibitory synapses (the release of GABA and glycine) may be more sensitive to tonic activity. GABARAP is an acronym for GABA associated receptor protein. GAD converts glutamate to GABA by decarboxylation (not shown)

antigens, on the other hand, have been proven in tests of live cultured neurons to not reach their intracellular targets and are hence not harmful. They are, however, valuable as biomarkers for these paraneoplastic illnesses, which are mostly caused by T cell pathology. This is reinforced by autopsy evidence that demonstrates cytotoxic T lymphocytes infiltrating neurons and triggering degeneration via perforin and granzyme B-mediated pathways. As a result, immunotherapeutic treatments involving the elimination of antibodies or antibody-producing cells are ineffective in patients with these paraneoplastic CNS illnesses (Avidan et al. 2014). Autoantibodies may be directed against neuronal cell surface and synaptic receptor proteins (Crisp et al. 2016). Autoantibodies binding to extracellular epitopes of neuronal cell surface proteins have been found to be secondary to the pathogenic effect of some recently reported encephalitides (such as NMDAR, LGI1, and CASPR2 encephalitis). Immunotherapy is successful in removing or neutralizing the effect of these pathogenic autoantibodies. Myasthenia gravis and Eaton-Lambert syndrome are caused by autoantibodies against cell surface proteins at the neuromuscular junction (acetylcholine receptors and voltage gated calcium channels). Immunotherapy has been proven to be useful in the treatment of various diseases.

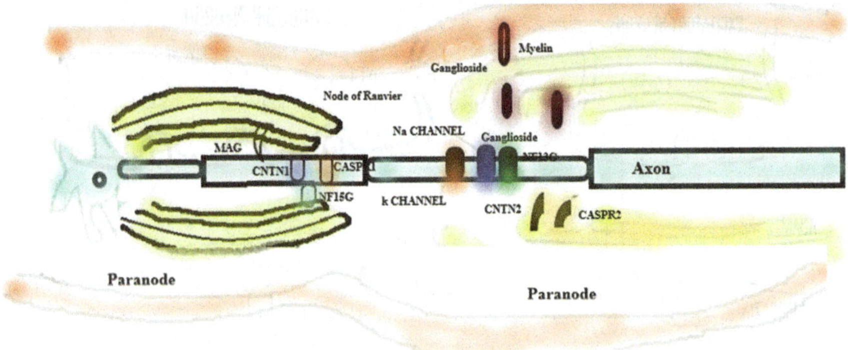

Fig. 5.3 Autoimmune diseases of peripheral nerve and neuromuscular junction – Peripheral nerve: Guillain Barre syndrome and its variants have been linked to autoantibodies to gangliosides (glycosphingolipids) which are abundant in peripheral nerves. Autoantibodies cause segmental demyelination and disruption of Na channel clustering. GM1 and GalNAc-GD1a ganglioside are most abundant in axolemma of motor nerves and nodes of Ranvier and are associated with motor phenotypes. GQ1b is enriched in oculomotor cranial nerves and is linked with Miller Fisher syndrome. Multifocal motor neuropathy (MMN) is linked to anti-GM1 ganglioside antibodies. MGUS neuropathy is linked to autoantibodies to myelin associated glycoprotein (MAG). Chronic inflammatory demyelinating polyneuropathy (CIDP) is linked to autoantibodies

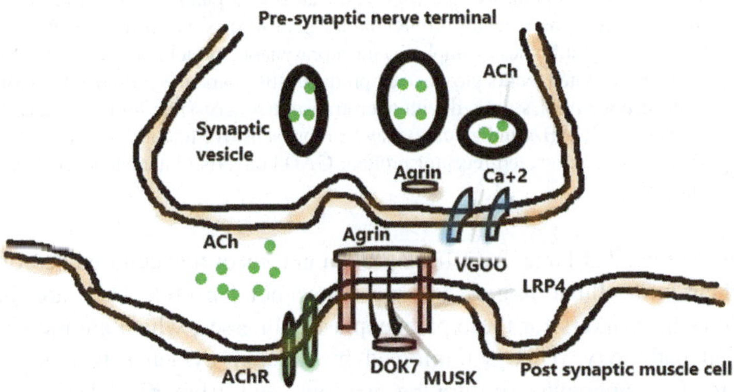

Fig. 5.4 Autoantibodies to the Aquaporin 4 (AQP4) water channel, which is found throughout the body, especially on astrocytic foot processes abutting capillary endothelial cells and ependymal cells, cause neuromyelitis optica (NMO). Pathology is most prevalent in the peri-ependymal regions of the spinal cord, optic nerve, and brain

Glutamic acid decarboxylase (GAD), an intracellular enzyme whose physiological function is the decarboxylation of glutamate to gamma-aminobutyric acid (GABA), the main inhibitory transmitter within neurons, illustrates a more complicated pattern of expression in addition to these two categories (intracellular or cell surface antigenic expression). GAD 65 can bind to the plasma membrane and then escape into the extracellular area. During synaptic fusion and reuptake, antibodies can be exposed to GAD65 and amphiphysin epitopes. Anti-GAD65 titers above a certain threshold have been linked to stiff person syndrome, cerebellar ataxia, and limbic encephalitis (Budharan et al. 2021).

5.10 Clinical Characteristics Triggering Autoimmune Diseases

The symptoms of autoimmune encephalitis are like those of encephalitis caused by infections. A lowered or changed level of awareness, tiredness, short-term memory loss, personality changes such as apathy, impatience, restlessness, and, most typically, convulsions are all signs of encephalitis. The following symptoms (which appear over days to weeks rather than years) should be looked for in people with probable encephalitis.

The causes leading to autoimmunity include:

- Acute or subacute cognitive deterioration, behavioral abnormalities, and seizures (usually intractable and resistant to anti-epileptic drugs) should urge evaluation of autoimmune encephalitis when viral causes have been ruled out. Antibodies against NMDAR, LGI1, and GABABR cause the greatest seizures in autoimmune encephalitis. When this clinical presentation is accompanied by an aberrant signal on MRI FLAIR images of the bilateral medial temporal lobe, limbic encephalitis should be evaluated. Although CNS diseases such as HSV or HHV6 encephalitis can induce limbic encephalitis, the most common antibodies associated with limbic encephalitis are anti-Hu, Ma2, LGI1, GABAB, or AMPA receptor.
- Faciobrachial dystonic seizures are characterized by lateralized (unilateral or bilaterally asynchronous) tonic contractions that primarily affect the upper limb and face and are frequently linked with hand dystonia. Anti-leucine-rich glioma inactivated (LGI1) encephalitis is characterized by recurrent episodes that might occur multiple times per day (Morano et al. 2020). Agitation, aggression, and violence, visual or auditory hallucinations, delusions, disorganized or bizarre behavior, paranoia, disinhibition, apathy, mutism, and catatonia; intolerance to neuroleptics may cause hyperthermia, rigidity, rhabdomyolysis, or coma; intolerance to neuroleptics may cause hyperthermia, rigidity, rhabdomyolysis, or coma. NMDAR encephalitis (which affects up to 95% of patients) is the most common cause of these symptoms (Pollack et al. 2020). Cerebellar ataxia (pan cerebellar involvement) that appears suddenly (within 3 months) or is accompanied with limbic or brain stem encephalitis, myelitis, or neuropathy. AntiHu (ANNA-1), mGluR1, CASPR2, GAD65, Ri (ANNA2), Yo (PCA1), PCA2, CV2/CRMP5, or Sox1 are some of the antibodies linked to it. Adult Opsoclonusmyoclonus in solo or in combination with cerebellar myoclonus Anti-Ri antibodies may be linked to ataxia. Bickerstaff brainstem encephalitis is characterized by a decreased degree of consciousness, bilateral external opthalmoplegia, and ataxia that develops over 4 weeks and is related with antibodies to anti-GQ1b antibodies (Bhagavati 2021).

Atypical sleep architecture with undifferentiated NREM sleep, REM sleep behavior disorder (RBD), obstructive sleep apnea and stridor, and agrypnia excitata are all examples of sleep disorders (insomnia, motor, and autonomic hyperactivation). Patients with antibodies to IgLON5, CASPR2, Ma2, and LGI1 have been documented as having a variety of sleep problems and RBD symptoms (Sabater et al. 2014). Secondary narcolepsy can be a symptom of NMOSD with hypothalamic involvement.

Various autoantibodies have been found to be associated with specific movement disorders, such as the characteristic dyskinesias, chorea, and dystonia associated with NMDAR antibodies, stiff person spectrum disorders associated with GAD, glycine receptor, amphiphysin, or DPPX antibodies, specific paroxysmal dystonias associated with LGI1 antibodies, and cerebellar ataxia associated with various anti-neuronal antibodies (Balint et al. 2018).

Peripheral nerve hyperexcitability (neuromyotonia) manifests as cramps, myokimia, or fasciculations, and is linked to encephalopathy (anti-CASPR2 and LGI1 antibodies). Antibodies that target proteins expressed by inhibitory synapses such as glutamic acid decarboxylase 65 (GAD65), subunit of the glycine receptor (GlyR), or amphiphysin (Crisp et al. 2019) are associated with stiff person spectrum disorders presenting as fluctuating muscle spasms and rigidity, as well as acquired hyperekplexia (startle response). Hyperthermia, blood pressure variations, tachycardia, less commonly bradycardia necessitating a temporary pacemaker, excessive salivation, hyperhidrosis, chronic gastro-intestinal pseudo-obstruction, and central hypoventilation are examples of autonomic abnormalities. Anti-NMDAR and anti-Hu encephalitis are the most common causes of these symptoms.

Sensory neuronopathies are a rare set of illnesses caused by dorsal root ganglion degeneration that manifest as asymmetric, multifocal, non-length dependent sensory loss, neuropathic pain, and subacute sensory ataxia. Anti-Hu, CV2/CRMP5, and amphiphysin antibodies have been related to this illness in cancer patients, either alone or in combination with encephalomyelitis. Multiple, massive white matter and deep gray matter demyelinating lesions on brain MRI scans reveal acute disseminated encephalomyelitis in children and young adults with multifocal encephalitic presentation (ADEM). In up to 50% of instances, MOG antibodies can be found in the blood. Antibodies to AQP4 (NMO spectrum illness) or MOG (MOG antibody-associated disease) or multiple sclerosis (typically unilateral) (Chen et al. 2020) may be associated with unilateral or bilateral optic neuritis manifesting as rapid visual loss. AQP4 antibodies (NMO spectrum illness), MOG antibodies (MOG antibody-associated disease), MS, or ADEM may be linked to myelitis appearing as rapid onset paraparesis or quadriparesis (Chen et al. 2020; Lopez-Chiriboga et al. 2018).

5.11 B Cells Play a Key Role in Nervous System Autoimmune Diseases

T and B lymphocytes both play a role in the pathophysiology of autoimmune inflammatory nervous system diseases (Weissert 2017). Recently, the importance of B lymphocytes in some of these diseases has been highlighted (Sabatino 2019).

An altered B cell receptor (BCR) allows B cells and their produced antibodies to identify different antigens. Hematopoietic stem cells in the bone marrow give rise to naive B cells, which circulate in the blood and lymphoid tissues. Following detection of their relevant antigen in germinal centers, a subset of CD27+ memory B cells may develop into CD38+ antibody-secreting plasma blasts and long-lived CD138+ plasma cells. B cells can internalize proteins and present peptide fragments to antigen specific CD8+ and CD4+ T cells in combination with MHC Class I and Class II molecules, regulating T cell activity in addition to antibody formation. Antigen presentation by B cells appears to play a major role in T cell activation in MS, according to current findings (Sabatino 2019). B cells also produce pro-inflammatory cytokines like lymphotoxin, interleukin 6, and tumor necrosis factor, as well as anti-inflammatory cytokines like IL-10 and IL-35, which can promote or reduce CNS inflammation. Changes in B cell development in the bone marrow because of altered B cell intrinsic signals have been linked to an altered naive B cell repertoire and the formation of autoantibody-producing B cells, according to new research (Rawlings et al. 2017). B cell activating factor (BAFFR), CD40, and Toll-like receptors regulate a complex interaction of BCR and co-receptor signaling in these signals (TLRs). As a result of altered signaling programs (because of genetic variations linked to autoimmunity), a larger proportion of mature B cells with autoreactivity may be produced (Rawlings et al. 2017).

5.12 CNS Autoimmune Disorders

5.12.1 Multiple Sclerosis (MS)

Although T cells have traditionally been thought to drive the autoimmune response to unknown target antigens in MS, the importance of B cells in pathophysiology has been highlighted by (1) the detection of CSF oligoclonal bands, which are unique IgG fractions produced by an inthrathecal clonal B cell population that target ubiquitous self-antigens; (2) B cells seen in MS CNS lesions, most commonly in active lesions; (3) ectopic lymphoid follicle-like structures containing CD20+ B cells, CD138+ plasma cells, and follicular dendritic cells found in the leptomeninges of MS patients., (4) the identification of CSF antibody reactivity against measles, rubella, and varicella-zoster viral antigens; (5) the expansion of populations of plasma blasts and plasma cells collected from MS patients, which target neurons, astrocytes, and oligodendrocytes, but the target antigen is unknown; (Jarius et al. 2017). This research implies that the humoral B cell response in MS may not be directed against a single antigen, but rather against a wide range of self and non-self-antigens that may vary from person to person (Sabatino 2019). The diverse set of targets could be due to epitope dissemination and secondary immune responses to CNS injury. The exceptional success of CD20-targeting treatments in MS patients has revealed B cells' crucial role in MS pathophysiology. Anti-CD20 B cell therapy has a clinical benefit in MS that comes before a reduction in total IgM and IgG levels, indicating that the impact is not related to a reduction in humoral immunity but rather to decreased antigen presentation to T cells or a decrease in inflammatory cytokines like IL-6 or TNF. Furthermore, the most effective MS treatments include the ablation or inhibition of B and T cell trafficking in the periphery, which leads to a significant reduction in new CNS lesions and relapses. This shows that BBB collapse and recruitment of peripheral lymphocytes and macrophages into the CNS are involved in MS relapses (Sabatino 2019). As a result, the lack of efficacy in some types of MS, such as progressive MS, could be due to the aberrant immune response being sequestered behind an intact blood-brain barrier.

5.12.2 Neuro-Myelitis Optica Spectrum Disease (NMOSD)

NMOSD is a demyelinating inflammatory lesion that primarily affects the optic nerves, spinal cord, and brainstem. Anti-Aquaporin 4 (a membrane-bound water channel extensively expressed on astrocytic foot processes) IgG is found in 70–90% of NMO patients. Anti-AQP4 antibodies have been shown to be harmful in vitro and in vivo, and intracerebral injection of AQP4 IgG with complement causes demyelination like NMOSD.

AntiAQP4 antibody titers in serum are 1000-fold higher than in CSF, and CSF oligoclonal bands, which are found in only 15–30% of NMO patients, frequently fade away with disease progression, showing that B cell activation and the origin of the humoral immune response are located outside the CNS. Following BBB breach, activated B lymphocytes may penetrate the CNS and produce disease, as previously stated. Tolerance is maintained during normal early B cell growth by the elimination of most self-reactive B cells. AntiAQP4 IgG-producing B cells appear to emerge from early B cell tolerance checkpoint abnormalities, both centrally and peripherally in the bone marrow, resulting in an increased number of autoreactive B cells in the mature naive B cell population (Cotzomi et al. 2019). After somatic hypermutations, this reservoir of auto-reactive B cells can provide B cell clones that produce pathogenic anti-AQP4 antibodies. The fact that unmutated precursors of B cells secreting anti-AQP4 antibodies do not bind to autoantigen supports the conclusion that anti-AQP4 specificity and the generation of pathogenic autoantibodies require affinity maturation and the acquisition of somatic hypermutations, both of which are regulated by T cells (Cotzomi et al. 2019). Additionally, B cell clones may be activated by an antigen other than AQP4, such as a peptide from Clostridium perfringens, an intestinal bacterium native to the United States (which has homology to an immunodominant epitope of AQP4).

5.12.3 Myelin Oligodendrocyte Glycoprotein (MOG) Antibody-Associated Disease

MOG is made by oligodendrocytes, the CNS's myelin-forming cells. On the surface of oligodendrocytes and the exterior lamella of myelin sheaths, it is expressed. It has traditionally piqued researchers' attention in the study of MS because it was shown to be the target antigen in an animal model of MS called experimental allergic encephalomyelitis (von Büdingen et al. 2001). MOG antibodies, on the other hand, were shown to be non-specific in MS investigations (Karni et al. 1999). MOG antibody has only recently emerged as a biomarker for CNS inflammatory demyelinating illnesses different from MS and NMOSD, thanks to improved antibody detection methods using cell-based assays that detect MOG in its original structural state. According to a recent study, the best reliable tests for detecting MOG antibodies are those that use live cells expressing native full length human MOG (since some conformational epitopes are lost when MOG expressing cells are fixed) (Reindl et al. 2020). Second, low positive samples are widely discovered, particularly in healthy people, necessitating more research towards a suitable cutoff value for identifying clinically relevant results. Because the reliability of CSF MOG-IgG is unknown, serum testing is indicated for detecting MOG-IgG. In contrast to MS, oligoclonal bands are rarely identified in CSF, showing that the autoimmune process begins in the peripheral nervous system. Prior to the onset of MOG-antibody linked sickness, prodromal symptoms such as fever, malaise, cough, and rhinorrhea are prevalent (Flanagan 2019). Optic neuritis, transverse myelitis, or acute disseminated encephalomyelitis (ADEM, which is most common in people under the age of 20) are the most common symptoms (Jurynczyk et al. 2017). Ataxia, facial palsy, diplopia, or vertigo may be signs of brain stem or cerebellar involvement (Banks et al. 2020). In addition, overlapping central and peripheral nervous system involvement (acute inflammatory demyelinating neuropathy, myeloradiculitis, multifocal motor neuropathy, or brachial neuritis) has been observed in a recent study (Rinaldi et al. 2021). A monophasic or recurrent clinical course is possible. High titers and chronic MOG-IgG positive in ADEM are associated with a higher likelihood of recurrence (Lopez-Chiriboga et al. 2018). Optic neuritis can be unilateral or bilateral, and it is usually accompanied with edema of the optic disc. The optic nerve, including the optic sheath, may exhibit amplification on an MRI. It is common to have bilateral anterior optic nerve enhancement without extension to the optic chiasm (Chen et al. 2020). Most individuals with myelitis on MRI have longitudinally significant lesions, with involvement of the conus found more frequently than in NMOSD. On MRI, lesions in the spinal cord usually affect the gray matter, which distinguishes them from lesions in MS and NMOSD (Dubey et al. 2019).

5.12.4 NMDAR Encephalitis

NMDAR encephalitis is the most prevalent autoimmune encephalitis, with IgG autoantibodies directed against the NMDAR's GluN1 subunit (Kreye et al. 2016). The disease has a female predominance (female to male ratio of 8:2) and is frequently linked with ovarian teratoma. The median age at presentation is 21 years (range 1–85 years). Abnormal behavior (visual or auditory hallucinations, acute schizoaffective disorder, depression, and mania) is the most common symptom, followed by cognitive dysfunction, seizures, movement disorders (oral, facial, and lingual dyskinesias, chorea, athetosis, and dystonia), autonomic dysfunction, and central hypoventilation. The presence of NMDAR IgG antibodies in the cerebrospinal fluid confirms the diagnosis. B cells, CD4+ T cells, and IgG deposits are found in infiltrates of B cells, CD4+ T cells, and IgG deposits in the brain, but there is little or no neuronal damage, explaining the considerable recovery that can be accomplished following early immunotherapy. NMDAR antibodies in patients' serum are harmful, according to multiple lines of evidence (Hughes et al. 2010). Antibodies crosslink NMDARs in cultured neurons, causing their internalization and impairment of synaptic plasticity (Hughes et al. 2010). Memory impairments, anhedonia, and seizures were observed in rats when patient CSF or IgG isolated from patient CSF was infused into the ventricular space via intraventricular catheters (Taraschenko et al. 2019). Furthermore, mice that are actively immunized with GluN1/GluN2B heteromers have a fulminant encephalitis (Jones et al. 2019). In cultured neurons, the NMDAR antibodies elicit a decrease in NMDAR and NMDAR currents. NMDAR antibodies are also transferred transplacentally from pregnant patients to embryos, resulting in neurological abnormalities in newborns (Shi et al. 2017). Overall, these findings confirm Witbesky's postulates that NMDAR autoantibodies play a role in illness pathogenesis. Anti-NMDAR antibodies are more reliably found in the CSF than in serum, in contrast to NMOSD (where antibodies are considerably more readily detectable in serum than in CSF) (both should be tested). In addition, patients with NMDAR encephalitis have clonally increased NMDAR-specific plasma cells in their CSF, and B cells in their CSF can manufacture anti-NR1 antibodies. In contrast to myasthenia gravis and NMOSD, in which antibody production occurs in the periphery, our findings show that the anti-NMDAR encephalitis NMDAR specific B cell response is compartmentalized in the CNS.

5.12.5 PNS Autoimmune Disorders

Myasthenia gravis (autoantibodies directed against the post-synaptic Acetylcholine receptor or muscle specific tyrosine kinase) and Lambert-Eaton syndrome (autoantibodies directed against pre-synaptic voltage gated calcium channels) are two autoimmune illnesses of the neuromuscular junction. One of the classic examples of an autoimmune disease is myasthenia gravis, which has been definitively proven to have an autoimmune etiology (Patrick and Lindstrom 1973; Toyka et al. 1975). Although the evidence is not conclusive, there is a lot of indirect evidence that various skeletal muscle and peripheral nerve illnesses have an autoimmune origin.

5.12.6 Myopathies Caused by the Immune System

Immune-mediated myopathies, which include dermatomyositis, necrotizing autoimmune myopathy, overlap myositis, and inclusion body myositis are a set of acquired muscle illnesses (Milone 2017; Selva-O'Callaghan et al. 2018). Except for inclusion body myositis, many illnesses respond well to immunosuppressive medication. Dermatomyositis is characterized by proximal muscle weakness and a rash on the skin. Perifascicular atrophy, perimysial and perivascular infiltrates of CD4+ T cells, B cells, and plasmacytoid dendritic cells, and sarcolemmal overexpression of MHC1 are seen in muscle biopsies. Dermatomysositis manifests early with the deposition of membrane assault complex on intramuscular capillaries, although the origin of the microangiopathy is unknown. Around 70% of people with dermatomyositis have a distinct autoantibody that is linked to specific clinical characteristics. Antibodies against anti-transcriptional factor 1 (TIF-1) or anti-nuclear matrix protein NXP2 are two examples. It is linked to a higher risk of cancer. Antibodies to the melanoma differentiation associated gene 5 (MDA5) cause significant cutaneous abnormalities such as skin ulcers and interstitial lung disease in patients. Antisynthetase syndrome is a common symptom of overlap myositis, which is a kind of autoimmune myopathy linked to other connective tissue diseases. Autoantibodies to several aminoacyl tRNA synthetases are present in these patients. Myopathy is related with interstitial lung disease, radial finger skin lesions known as mechanic's hands, and Raynaud's phenomenon in patients with anti-Jo antibodies. Proximal muscular weakness, high creatine kinase levels, and muscle biopsies demonstrating necrosis and regeneration with low or no lymphocytic infiltration, overexpression of MHC1, and membrane-attack complex deposition on non-necrotic fibers are all symptoms of immune-mediated necrotizing myopathy. Autoantibodies to signal recognition particle (anti-SRP) or 3-hydroxyl 3-methylglutaryl coenzyme-A reductase (HMGCR), an enzyme that catalyzes the rate-limiting step in cholesterol production, are found in about two-thirds of patients. Antibodies to HMGCR are most found in persons who have taken statins. On muscle biopsies, patients with inclusion body myositis have CD8+ inflammatory infiltrates as well as rimmed vacuoles, ragged red fibers indicating mitochondrial damage, and aberrant protein aggregates, indicating both an inflammatory and a degenerative process. In 30–60% of patients with sporadic inclusion body myositis, autoantibodies against cytosolic 5′-nucleotidase 1 A are detected.

5.12.7 Immune-Mediated Neuropathies

Acute inflammatory demyelinating polyradiculoneuropathy (AIDP; Guillain-Barre syndrome), chronic inflammatory demyelinating polyneuropathy (CIDP), multifocal motor neuropathy (MMN), and neuropathies associated with IgM monoclonal gammopathy of unknown significance (MGUS) are examples of immune-mediated neuropathies (Querol et al. 2017). Antibodies directed against myelin antigens, as well as autoreactive T cells and macrophages that penetrate the myelin sheath or axonal membranes, are thought to be involved in autoimmune peripheral nerve injury. Autopsies of patients with Guillain-Barre syndrome revealed perivascular and endoneurial inflammatory infiltrates of T cells and macrophages throughout the nerves, roots, and plexuses, as well as segmental demyelination caused by macrophages (Yuki and Hartung 2012). Autoantibodies to various gangliosides, which are abundant in axons, are found frequently in patients with the Miller-Fisher subtype (GQ1b, GT1a) or the axonal variant (GM1, GD1a) of Guillain-Barre syndrome (80–90%), but infrequently in patients with the most common demyelinating subtype (GM1, GD1a). Autoantibodies to numerous proteins found at the nodal or paranodal areas of peripheral nerves (such as CASPR1, contactin1, and NF155) that maintain axon-Schwann cell or axon-myelin binding have been found in 15–20% of CIDP patients' serum. These autoantibodies may cause axonal degeneration, demyelination, and alteration of nodal architecture by activating complement and macrophages. Sensory neuropathies caused by dorsal root ganglion pathology linked to anti-Hu, CRMP5, or amphiphysin antibodies are known as paraneoplastic neuropathies. Antibodies to CASPR2 can cause peripheral nerve hyperexcitability and neuropathic pain. Anti-Hu antibodies have been linked to enteric autonomic involvement and severe gastrointestinal mobility disorders. Plasmapheresis, IVIG, and/or immunosuppression are commonly used to treat these illnesses.

5.12.8 Autoimmunity and Liver Disease

Autoimmune liver diseases (AILDs) are distinguished by progressive immune-mediated inflammation and eventual destruction of hepatocytes and biliary epithelial cells (Barba Bernal et al. 2021). It primarily includes autoimmune hepatitis (AIH), primary biliary cholangitis (PBC), and primary sclerosing cholangitis (PSC), though overlap syndromes may occur (Invernizzi and Mackay 2008). Immunoglobulin G (IgG4)-related hepatobiliary diseases, such as IgG4-sclerosing cholangitis and IgG4-autoimmune hepatitis, account for a small proportion of AILDs (Montano-Loza et al. 2017). Liver transplantation (LT) is a treatment option for patients who have irreversible liver damage. With a few exceptions, the indications for LT in patients with AILDs are like those for patients with other chronic liver diseases (Ilyas et al. 2011). These autoimmune hepatobiliary diseases account for roughly 24% of total LT, making them the third most common LT indication in most transplant centers (Liberal et al. 2013). Each autoimmune liver disease necessitates a unique therapeutic approach, but the goals of treatment are the same: to improve the recipient's survival, prevent liver graft failure, and reduce disease recurrence. To achieve these goals, several factors must be considered, including identifying post-LT risk factors that correlate with liver graft failure and disease recurrence, selecting the most appropriate immunosuppressive regimen, and implementing additional cancer surveillance depending on the LT indication (Ilyas et al. 2011).

AIH is a complex inflammatory liver disease with an unknown cause that is caused by a combination of immunologic, genetic, and environmental factors. The prevalence of AIH among adults varies greatly around the world, ranging from 4 (Singapore) to 42.9 (Alaska) cases per 100,000 people (Czaja 2017). A recent population-based national study in the United States reported an estimated prevalence of 31.2 cases per 100,000 people (Tunio et al. 2021). The clinical presentation is quite variable. Patients can be asymptomatic, have acute symptoms (acute severe or acute liver failure), or have mild symptoms that can become chronic. The diagnosis is difficult and necessitates specific laboratory indicators such as elevated aminotransferase levels and increased total Immunoglobulin G; lymphoplasmacytic interface hepatitis on histology, and the presence of autoantibodies in the serum (antinuclear antibodies, anti-smooth muscle antibody, and anti-liver kidney microsomal antibodies). Other liver diseases, such as viral hepatitis, Wilson's disease, hereditary or metabolic liver injuries, should be ruled out (Mack et al. 2020).

Despite regular changes to AIH management guidelines, a considerable number of patients with AIH advance to end-stage liver disease and eventually require an LT. AIH accounts for 5% of all LT procedures performed in the United States (Ilyas et al. 2011). At 5 years, patient and graft survival rates are 76% and 70.9%, respectively (Suri et al. 2020). These rates are rather excellent; nevertheless, when compared to other LT indications, there is still a higher risk of late acute rejection (9%), chronic rejection (16%), and recurrent illness (36–68% at 5 years) (Stirnimann et al. 2019). There are difficulties in diagnosing recurrent AIH (rAIH). It is tough to tell the difference between rAIH and graft rejection because of factors including immunosuppressive medication and the disease's brief duration. The characteristics of the biopsy aid in the differentiation of these two conditions (Suri et al. 2020). The most important aspect of post-LT care for AIH is determining which immunosuppressive maintenance therapy is most appropriate. To balance their possible impact on the long-term morbidity and mortality of AIH patients after LT, transplant hospitals are adopting new immunosuppressive regimes. Post-LT results in patients with AIH have been documented in several systematic reviews and meta-analyses (Gautam et al. 2006; Montano-Loza et al. 2017). However, management findings remain restricted, making it difficult to make management suggestions based on high-quality information.

Autoimmune hepatitis (AIH) is an immunoinflammatory chronic liver disease that manifests itself in a variety of ways. Worldwide, there has been an increase in the prevalence of AIH, as well as a proportional increase in the percentage of male patients. Serum biochemistry and liver histology are used to diagnosis AIH, which includes increased aminotransferases and serum immunoglobulin G (IgG), serum anti-nuclear antibody or anti-smooth muscle antibody, and interface lymphoplasmacytic hepatitis. Clinical symptoms range between disease subtypes with different time frames, such as AIH with a chronic gradual onset and AIH with an abrupt onset (the diagnosis of which is often challenging due to the lack of typical serum findings). Because there are no disease-specific biomarkers or histological evidence, the disease phenotype may be expanded to include drug-induced AIH-like liver injury. The first-line therapy for AIH include corticosteroids and azathioprine. To achieve long-term overall survival, full normalization of aminotransferases and serum IgG is a critical therapy response. Second-line treatment is recommended when these medications produce a partial response or intolerance, notably with mycophenolate mofetil. The majority of patients will require lifelong maintenance treatment; however, those who achieve extended and strict biochemical remission with normalized alanine aminotransferase and IgG levels may be able to stop taking the drugs. In the future, customized medicine should be used to manage the quality of life of AIH patients, including the proper selection and dose of first-line therapy, maybe alternating with prospective therapies, and the prediction of treatment withdrawal success (Komori 2021).

5.12.9 Rheumatoid Arthritis

Rheumatoid arthritis (RA) is a chronic inflammatory illness that causes joint swelling, pain, and synovial joint degeneration, resulting in severe disability. RA is classified as an autoimmune disorder (Aletaha et al. 2010). According to a study conducted in the United Kingdom, the prevalence of RA in the general population is 1.16% in women and 0.44% in men (Symmons et al. 2002).

Rheumatoid arthritis (RA) is an inflammatory illness that causes inflammation in the joints. The presence of autoantibodies in RA patients' serum has revealed a lot about the disease's biology. RA can be separated into seropositive and seronegative illness based on the presence of autoantibodies such as rheumatoid factor (RF), anti-citrullinated protein antibodies (ACPA), anti-carbamylated protein antibodies (anti-CarP), and more recently anti-acetylated protein antibodies. Specific human leukocyte antigen (HLA) alleles and smoking are linked to the development of these autoantibodies, which are linked to both genetic and environmental risk factors for RA. In a subset of individuals, autoantibodies can be found many years before disease manifestation, implying a series of events in which the initial autoantibodies develop in susceptible hosts, followed by an inflammatory response and clinically evident arthritis. The features and effector roles of these autoantibodies are being studied in order to gain a better understanding of the pathophysiological processes that underpin arthritis in RA. Recent research reveals that ACPA may play a role in the persistence of inflammation once it has begun. Furthermore, pathophysiological mechanisms establishing a direct relationship between the presence of ACPA and both bone erosions and pain in RA patients have been established. Finally, examining potentially harmful autoantibodies' could lead to a better understanding of the underlying pathophysiological mechanisms in rheumatoid arthritis (Derksen et al. 2017).

Rheumatoid arthritis (RA) is a chronic autoimmune disease that affects the joints principally. RA is a diverse disease with numerous disease subtypes that have likely different underlying etiology. These distinct pathophysiological mechanisms may lead to a similar clinical appearance of arthritis via a final shared inflammatory pathway. The most well-known RA subtypes are ACPA-positive and ACPA-negative illness, which differ in risk factors and clinical outcomes (Scott et al. 2010). Several autoantibodies can be seen in the serum of RA patients, the most prominent of which are rheumatoid factor (RF) and anti-citrullinated protein antibodies (ACPA). Antibodies against additional posttranslationally changed proteins, such as anti-carbamylated protein antibodies (anti-CarP) (Shi et al. 2011) and anti-acetylated protein antibodies (anti-ACP) (Juarez et al. 2016), have recently been found. Most of the recent research on the role of autoantibodies in disease pathophysiology has focused on ACPA, which are anti-citrullinated proteins.

Peptidyl-arginine deiminase (PAD) enzymes catalyze citrullination, which converts the DNA-encoded amino acid arginine to citrulline. Both healthy and pathological conditions result in posttranslational alteration. The development of the immune response against citrullinated proteins, and consequently the production of ACPA, is thought to be linked to several known risk factors for RA.

5.12.9.1 Autoantibodies in RA

Autoantibodies are thought to be present in 50–80% of RA patients (Scott et al. 2010). As previously stated, the existence of autoantibodies has allowed for the identification of RA subgroups that are more homogeneous in terms of risk factors as well as clinical disease history. The first autoantibody system described in RA was RF, an autoantibody directed against the Fc portion of human IgG. Despite its low specificity, the presence of RF was regarded so typical of RA that it was included in the 1987 ACR classification criteria for RA. The more RA-specific ACPA was identified some decades later (Derksen et al. 2017). Both RF and ACPA are included in the ACR-EULAR 2010 classification criteria for RA. Antibodies to other posttranslationally changed proteins, such as carbamylated (Shi et al. 2011) and acetylated proteins, have recently been discovered (Juarez et al. 2016). Seropositive RA is linked to faster radiographic development and joint destruction (van der Helm-van Mil et al. 2005), whereas seronegative RA patients present with greater inflammatory markers (Nordberg et al. 2017). Furthermore, not only is the presence of a single autoantibody essential for identifying various phenotypes of RA patients, but the presence of several autoantibodies may be as well (Derksen et al. 2016). Autoantibodies not only provide information on disease outcome, but they also provide insight into how RA develops. A better knowledge of the underlying pathophysiological mechanisms in rheumatoid arthritis has resulted from research into the various autoantibodies and their properties (Derksen et al. 2017).

5.12.10 Anti-Citrullinated Protein Antibodies

Citrullinated peptides are produced because of a posttranslational alteration mediated by PAD enzymes, as previously stated. Antibodies against citrullinated peptides of various isotypes, including IgG, IgA, and IgM, have been seen in RA (Verpoort et al. 2006). Because IgA is linked to a mucosal origin of the immune response, the presence of ACPA IgA supports the concept that ACPA is linked to smoking or microbiome dysbiosis. Citrullinated proteins are seen in synovial fluid from inflamed RA joints, suggesting that ACPA could bind to these antigens in the joint and enhance local inflammation (van Beers et al. 2013). Vimentin is a potential ACPA target protein. Passive transfer of ACPA in mice models of collagen-induced arthritis (CIA) does not create synovitis, but it can aggravate pre-existing synovitis (Kuhn et al. 2006). As a result, it's thought that numerous "hits" are required for the onset of RA. Autoantibodies, according to one theory, may cause the non-resolution and chronicity of a typically transient immune response, such as following trauma or infection.

5.12.11 Anti-Carbamylated Protein Antibodies

Anti-carbamylated protein (anti-CarP) antibodies are a type of anti-posttranslationally modified protein antibody (AMPA) that has been found in RA patients. Carbamylation is a cyanide-mediated chemical process that converts a lysine to a homocitrulline. Renal illness, smoking, and inflammation, for example, can all raise cyanide levels and consequently carbamylation (Wang et al. 2007). Increased carbamylation, like citrullination, does not appear to be sufficient to break tolerance and trigger autoimmunity. Anti-CarP antibodies are found in only 12% of patients with renal impairment, compared to 44% of RA patients (Verheul et al. 2016).

Although a recent study found that smoking widened the immune response against carbamylated vimentin in mice models, epidemiological studies found no link between anti-CarP and smoking in RA patients (Ospelt et al. 2017). The relevance of smoking and other (environmental or genetic) factors in breaking carbamylated protein tolerance, such as fibrinogen, is still unknown. Although homocitrulline and citrulline have similar chemical structures, ACPA and anti-CarP are different autoantibody classes, with anti-CarP found in both ACPA-positive and ACPA-negative patients (Shi et al. 2011).

Furthermore, among patients who do not have RF or ACPA, anti-CarP is linked to radiographic advancement. Anti-CarP testing did not help the diagnostic classification of RA patients, as RF and ACPA are already significant predictors of illness (Ajeganova et al. 2017). Fetal calf serum (FCS), which contains a variety of carbamylated proteins, is commonly used in anti-CarP assays. It is still unclear which autoantigens anti-CarP binds to in vivo.

5.12.12 Anti-Acetylated Protein Antibodies

Anti-acetylated protein antibodies, which have been reported in roughly 40% of RA patients, primarily in the ACPA-positive group, are the most recent addition to AMPAs in RA patients. Because detection rates in seronegative RA patients were equivalent to those in patients with resolving arthritis, these antibodies are unlikely to become a new biomarker for detecting RA (Juarez et al. 2016). Anti-acetylated protein antibodies, on the other hand, may bring significant new insights into pathophysiology, particularly in an era where the microbiome appears to be becoming increasingly relevant. Although the underlying mechanism is unknown, acetylation is an enzyme process that can be influenced by bacteria. Anti-acetylated antibodies may thus establish a novel link between microbiome dysbiosis and the onset of autoimmunity in RA (Simon et al. 2012).

5.13 Pathogenic Potential of Autoantibodies

The growth of the ACPA repertoire before disease start and the relationship of autoantibodies with radiographic progression imply that the anti-citrulline immune response may play a role in disease pathophysiology. Furthermore, rituximab-mediated B cell reduction is efficacious in RA patients, with increased efficacy in ACPA- and RF-positive cases, implying that B cells (and maybe the autoantibodies they manufacture) play a role in disease pathophysiology (Cambridge et al. 2003). Autoantibodies appear to play a pathogenic role in RA, according to many lines of evidence. In Fig. 1.1, the model on the possible role of autoantibodies in disease pathophysiology of RA, as discussed in this review, is depicted.

5.14 Binding to Fc Receptors

In general, antibodies exert an influence on other cells via Fc receptor binding. Immune complexes containing ACPA and citrullinated fibrinogen, like recall antigen immune complexes (ICs), can induce TNF production by stimulating Fc receptors on macrophages (Laurent et al. 2011). The presence of RF-IgM or RF-IgA, which amplifies the Fc receptor-mediated immune response and increases complement activation, can alter the effector actions of ACPA ICs (Anquetil et al. 2015). This shows that ACPA and RF may play a synergistic role in RA pathogenesis, which is confirmed by epidemiological studies that show that combining ACPA and RF is linked to increased disease activity (Sokolove et al. 2014).

5.15 Complement Activation

Complement activation is another important effector action of antibodies. The classical pathway (started by C1q), the alternative pathway (initiated by C3), and the lectin pathway (initiated by mannose-binding lectin (MBL)) are the three mechanisms via which the complement system can be activated. Opsonization, membrane attack complex production, and chemotaxis Complement levels are lower in RA patients' synovial fluid, but complement cleavage products are higher, indicating accelerated complement activation. The ability of ACPA to recruit complement via both the classical and alternative pathways has been demonstrated, but not via the lectin pathway (Trouw et al. 2009). Taken together, the evidence suggests that ACPA may enhance the immune response in RA by binding to Fc receptors and activating complement.

References

Ajeganova S, van Steenbergen HW, Verheul MK et al (2017) The association between anti-carbamylated protein (anti-CarP) antibodies and radiographic progression in early rheumatoid arthritis: a study exploring replication and the added value to ACPA and rheumatoid factor. Ann Rheum Dis 76(1):112–118. https://doi.org/10.1136/annrheumdis-2015-208870

Aletaha D, Neogi T, Silman AJ et al (2010) Rheumatoid arthritis classification criteria: an American College of Rheumatology/European League Against Rheumatism collaborative initiative. Arthritis Rheum 62:2569–2581

Ali M, Boosi Narayana Rao K, Majumder P, Sarkar R, Mapa K (2020) Alterations in inter-organelle crosstalk and Ca (2+) signaling through mitochondria during proteotoxic stresses. Mitochondrion 57:37–46

Amedei A, Bergman MP, Appelmelk B, Azzurri A, Benagiano M, Tamburini C, van der Zee R, Telford JL, Vandenbroucke-Grauls CMJE, D'Elios MM et al (2003) Molecular mimicry between Helicobacter pylori antigens and H+K + -adenotriphosphatase in human gastric autoimmunity. J Exp Med 198:1147–1156

Anquetil F, Clavel C, Offer G et al (2015) IgM and IgA rheumatoid factors purified from rheumatoid arthritis sera boost the Fc receptor rand complement-dependent effector functions of the disease specific anti-citrullinated protein autoantibodies. J Immunol 194(8):3664–3674. https://doi.org/10.4049/jimmunol.1402334

Atkinson MA, Bluestone JA, Eisenbarth GS, Hebrok M, Herold KC, Accili D, Pietropaolo M, Arvan PR, Von Herrath M, Markel DS et al (2011) How does type 1 diabetes develop? The notion of homicide or Beta-cell suicide revisited. Diabetes 60:1370–1379

Avidan N, Le Panse R, Berrih-Aknin S, Miller A (2014) Genetic basis of myasthenia gravis-a comprehensive review. J Autoimmun 52:146–153. https://doi.org/10.1016/j.jaut.2013.12.001

Balint B, Vincent A, Meinck H-M, Irani SR, Bhatia KP (2018) Movement disorders with neuronal antibodies: asyndromic approach, genetic parallels and pathophysiology. Brain 141:13–36. https://doi.org/10.1093/brain/awx189

Banks SA, Morris PP, Chen JJ, Pittock SJ, Sechi E, Kunchok A et al (2020) Brainstem and cerebellar involvement in MOG0IgG-associated disorder versus aquaporin-4-IgG and MS. J Neurol Neurosurg Psychiatry. https://doi.org/10.1136/jnnp-2020-325121. [Epub ahead of print]

Barba Bernal R, Medina-Morales E, Goyes D, Patwardhan V, Bonder A (2021) Management of autoimmune liver diseases after liver transplantation. Transplantology 2:162–182. https://doi.org/10.3390/transplantology2020016

Barbarroja N et al (2012) Progression from high insulin resistance to type 2 diabetes does not entail additional visceral adipose tissue inflammation. PLoS One 7:e48155

Barcellini W, Fattizzo B (2020) The changing landscape of autoimmune hemolytic anemia. Front Immunol 11:946

Barcellini W, Giannotta J, Fattizzo B (2020a) Autoimmune hemolytic anemia in adults: primary risk factors and diagnostic procedures. Expert Rev Hematol 13:585–597

Barcellini W, Zaninoni A, Giannotta JA, Fattizzo B (2020b) New insights in autoimmune hemolytic anemia: from pathogenesis to therapy. J Clin Med 9:3859

Barcellini W, Giannotta JA, Fattizzo B (2021) Autoimmune complications in hematologic neoplasms. Cancers 13:1532. https://doi.org/10.3390/cancers13071532

Baumel-Alterzon S, Katz LS, Brill G, Garcia-Ocana A, Scott DK (2021) Nrf2: the master and captain of beta cell fate. Trends Endocrinol Metab 32:7–19

Beck RW, Bergenstal RM, Laffel LM, Pickup JC (2019) Advances in technology for management of type 1 diabetes. Lancet 394(10205):1265–1273. https://doi.org/10.1016/s0140-6736(19)31142-0

van Beers JJ, Schwarte CM, Stammen-Vogelzangs J, Oosterink E, Božič B, Pruijn GJ (2013) The rheumatoid arthritis synovial fluid citrullinome reveals novel citrullinated epitopes in apolipoprotein E, myeloid nuclear differentiation antigen, and beta-actin. Arthritis Rheum 65(1):69–80. https://doi.org/10.1002/art.37720

Betterle C, Presotto F (2008) Autoimmune polyendocrine syndromes (APS) or multiple autoimmune syndromes (MAS). In: Walker SA, Jara LJ (eds) Handbook of systemic autoimmune diseases. Endocrine manifestations of systemic autoimmune diseases. Elsevier Science, Amsterdam, pp 135–148

Bhagavati S (2021) Autoimmune disorders of the nervous system: pathophysiology, clinical features, and therapy. Front Neurol 12:664664. https://doi.org/10.3389/fneur.2021.664664

Bingol B, Sheng M (2016) Mechanisms of mitophagy: Pink1, Parkin, Usp30 and beyond. Free Radic Biol Med 100:210–222

Bizzaro N, Antico A (2014) Diagnosis and classification of pernicious anemia. Autoimmun Rev 13: 565–568

Bizzaro N, Antico A, Villalta D (2018) Autoimmunity and gastric cancer. Int J Mol Sci 19(2):377. https://doi.org/10.3390/ijms19020377

Bock FJ, Tait SWG (2020) Mitochondria as multifaceted regulators of cell death. Nat Rev Mol Cell Biol 21:85–100

Bradford-Hill A (1965) The environment and disease: association or causation? Proc R Soc Med 58:295–300

Brooks-Worrell B, Palmer JP (2012) Immunology in the Clinic Review Series; focus on metabolic diseases: development of islet autoimmune disease in type 2 diabetes patients: potential sequelae of chronic inflammation. Clin Exp Immunol 167:40–46

Brooks-Worrell BM, Juneja R, Minokadeh A, Greenbaum CJ, Palmer JP (1999) Cellular immune responses to human islet proteins in antibody-positive type 2 diabetic patients. Diabetes 48:983–988

Brooks-Worrell B, Narla R, Palmer JP (2012) Biomarkers and immune-modulating therapies for type 2 diabetes. Trends Immunol 33:546–553

Budharan A, Sechi E, Flanagan EP, Dubey D, Zekeridou A, Shah SS et al (2021) Clinical spectrum of high-titre GAD antibodies. J Neurol Neurosurg Psychiatry. https://doi.org/10.1136/jnnp-2020-325275. [Epub ahead of print]

von Büdingen HC, Tanuma N, Villoslada P, Villoslada P, Ouallet JC, Hauser SL et al (2001) Immune responses against the myelin/oligodendrocyte glycoprotein in experimental autoimmune demyelination. J Clin Immunol 21:155–170. https://doi.org/10.1023/A:1011031014433

Calay ES, Hotamisligil GS (2013) Turning off the inflammatory, but not the metabolic, flames. Nat Med 19:265–267

Cambridge G, Leandro MJ, Edwards JC et al (2003) Serologic changes following B lymphocyte depletion therapy for rheumatoid arthritis. Arthritis Rheum 48(8):2146–2154. https://doi.org/10.1002/art.11181

Chen J, Chernatynskaya AV, Li JW, Kimbrell MR, Cassidy RJ, Perry DJ, Muir AB, Atkinson MA, Brusko TM, Mathews CE (2017) T cells display mitochondria hyperpolarization in human type 1 diabetes. Sci Rep 7:10835

Chen J, Stimpson SE, Fernandez-Bueno GA, Mathews CE (2018) Mitochondrial reactive oxygen species and type 1 diabetes. Antioxid Redox Signal 29:1361–1372

Chen J, Pittock SJ, Flanagan EP, Lennon VA, Bhatti MT (2020) Optic neuritis in the era of biomarkers. Surv Opthalmol 65:12–17. https://doi.org/10.1016/j.survophthal.2019.08.001

Christen U, von Herrath MG (2011) Do viral infections protect from or enhance type 1 diabetes and how can we tell the difference? Cell Mol Immunol 8(3):193–198. https://doi.org/10.1038/cmi.2010.71

Cnop M et al (2005) Mechanisms of pancreatic β-cell death in type 1 and type 2 diabetes: many differences, few similarities. Diabetes 54(suppl 2):S97–S107

Cnop M, Foufelle F, Velloso LA (2012) Endoplasmic reticulum stress, obesity and diabetes. Trends Mol Med 18:59–68

Cotzomi E, Stathopoulos P, Lee CS, Ritchie AM, Soltys JN, Delmotte FR et al (2019) Early B cell tolerance defects in neuromyelitis optica favor anti-AQP4 autoantibody production. Brain 142: 1598–1615. https://doi.org/10.1093/brain/awz106

Crisp SJ, Kullmann DM, Vincent A (2016) Autoimmune synaptopathies. Nat Rev Neurosci 17: 103–117. https://doi.org/10.1038/nrn.2015.27

Crisp S, Dixon CL, Jacobson L, Chabrol E, Irani SR, Leite MI et al (2019) Glycine receptor autoantibodies disrupt inhibitory neurotransmission. Brain 142:3398–3410. https://doi.org/10.1093/brain/awz297

Czaja AJ (2017) Global disparities and their implications in the occurrence and outcome of autoimmune hepatitis. Dig Dis Sci 62:2277–2292

Danaei G et al (2011) National, regional, and global trends in fasting plasma glucose and diabetes prevalence since 1980: systematic analysis of health examination surveys and epidemiological studies with 370 country-years and 2.7 million participants. Lancet 378:31–40

De Block CE, de Leeuw IH, Bogers JJ, Pelckmans PA, Ieven M, van Marck EA, van Hoof V, Máday E, van Acker KL, van Gaal LF (2002) Helicobacter pylori, parietal cell antibodies and autoimmune gastropathy in type 1 diabetes mellitus. Aliment Pharmacol Ther 16:281–289

De Block CEM, de Leeuw IH, van Gaal LF (2008) Autoimmune gastritis in type 1 diabetes: a clinically oriented review. J Clin Endocrinol Metab 93:363–371

Derksen VF, Ajeganova S, Trouw LA et al (2016) Rheumatoid arthritis phenotype at presentation differs depending on the number of autoantibodies present. Ann Rheum Dis 76(4):716–720. https://doi.org/10.1136/annrheumdis-2016-209794

Derksen V, Huizinga T, van der Woude D (2017) The role of autoantibodies in the pathophysiology of rheumatoid arthritis. Semin Immunopathol 39(4):437–446. https://doi.org/10.1007/s00281-017-0627-z

Dixon MF, Genta RM, Yardley JH, Correa P (1996) Classification and grading of gastritis. The updated Sydney System. International Workshop on the Histopathology of Gastritis, Houston 1994. Am J Surg Pathol 20:1161–1181

Dubey D, Pittock SJ, Flanagan EP (2019) Clinical, radiographic and prognostic features of myelitis associated with myelin oligodendrocyte glycoprotein autoantibody. JAMA Neurol 76:301–309. https://doi.org/10.1001/jamaneurol.2018.4053

Dupuis J et al (2010) New genetic loci implicated in fasting glucose homeostasis and their impact on type 2 diabetes risk. Nat Genet 42:105–116

Eizirik DL, Miani M, Cardozo AK (2013) Signalling danger: endoplasmic reticulum stress and the unfolded protein response in pancreatic islet inflammation. Diabetologia 56:234–241

Eizirik DL, Pasquali L, Cnop M (2020) Pancreatic beta-cells in Type 1 and Type 2 diabetes mellitus: different pathways to failure. Nat Rev Endocrinol 16:349–362

Faller G, Kirchner T (2005) Immunological and morphogenic basis of gastric mucosa atrophy and metaplasia. Virchows Arch 446:1–9

Fattizzo B, Barcellini W (2020) Autoimmune cytopenias in chronic lymphocytic leukemia: focus on molecular aspects. Front Oncol 9:1435

Feuerer M et al (2009) Lean, but not obese, fat is enriched for a unique population of regulatory T cells that affect metabolic parameters. Nat Med 15:930–939

Filippi C, von Herrath M (2005) How viral infections affect the autoimmune process leading to type 1 diabetes. Cell Immunol 233(2):125–132. https://doi.org/10.1016/j.cellimm.2005.04.009

Flanagan EP (2019) Neuromyelitis optica spectrum disorder and other nonmultiple sclerosis central nervous system inflammatory diseases. Continuum 25:815–844. https://doi.org/10.1212/CON.0000000000000742

Gautam M, Cheruvattath R, Balan V (2006) Recurrence of autoimmune liver disease after liver transplantation: a systematic review. Liver Transpl 12:1813–1824

Gregor MF, Hotamisligil GS (2011) Inflammatory mechanisms in obesity. Annu Rev Immunol 29: 415–445

Guarda G et al (2009) T cells dampen innate immune responses through inhibition of NLRP1 and NLRP3 inflammasomes. Nature 460:269–273

van der Helm-van Mil AH, Verpoort KN, Breedveld FC et al (2005) Antibodies to citrullinated proteins and differences in clinical progression of rheumatoid arthritis. Arthritis Res Ther 7(5): R949–R958. https://doi.org/10.1186/ar1767

Hemminki K, Liu X, Ji J, Sundquist J, Sundquist K (2012) Effect of autoimmune diseases on mortality and survival in subsequent digestive tract cancers. Ann Oncol 23:2179–2184

Hughes EG, Peng X, Gleichman AJ, Lai M, Zhou L, Tsou R et al (2010) Cellular and synaptic mechanisms of anti-NMDA receptor encephalitis. J Neurosci 28:5866–5875. https://doi.org/10.1523/JNEUROSCI.0167-10.2010

Hundhausen C, Roth A, Whalen E et al (2016) Enhanced T cell responses to IL-6 in type 1 diabetes are associated with early clinical disease and increased IL-6 receptor expression. Sci Transl Med 8(356):356ra119. https://doi.org/10.1126/scitranslmed.aad9943

Igoillo-Esteve M et al (2010) Palmitate induces a proinflammatory response in human pancreatic islets that mimics CCL2 expression by β cells in type 2 diabetes. Diabetologia 53:1395–1405

Ilyas JA, O'Mahony CA, Vierling JM (2011) Liver transplantation in autoimmune liver diseases. Best Pract Res Clin Gastroenterol 25:765–782

Invernizzi P, Mackay I-R (2008) Autoimmune liver diseases. World J Gastroenterol 14:3290–3291

Jarius S, Eichhorn P, Franciotta D, Petereit HF, Akman-Demir G, Wick M et al (2017) The MRZ reaction as a highly specific marker of multiple sclerosis; reevaluation and structured review of the literature. J Neurol 264:453–464. https://doi.org/10.1007/s00415-016-8360-4

Jones BE, Tovar KR, Goehring A, Jalali-Yazdi F, Okada NJ, Gouaux E et al (2019) Autoimmune receptor encephalitis in mice induced by active immunization with conformationally stabilized holoreceptors. Sci Transl Med 11:eeaaw0044. https://doi.org/10.1126/scitranslmed.aaw0044

Juarez M, Bang H, Hammar F et al (2016) Identification of novel antiacetylated vimentin antibodies in patients with early inflammatory arthritis. Ann Rheum Dis 75(6):1099–1107. https://doi.org/10.1136/annrheumdis-2014-206785

Jurynczyk M, Messina S, Woodhall MR, Raza N, Everett R, Roca-Fernandez A et al (2017) Clinical presentation and prognosis in MOG-antibody disease: a UK study. Brain 140:3128–3138. https://doi.org/10.1093/brain/awx276

Kang S, Tanaka T, Narazaki M, Kishimoto T (2019) Targeting interleukin-6 signaling in clinic. Immunity 50(4):1007–1023. https://doi.org/10.1016/j.immuni.2019.03.026

Karni A, Bakimer-Kleiner R, Abramsky O, Ben-Nun A (1999) Elevated levels of antibody to myelin oligodendrocyte glycoprotein is not specific for patients with multiple sclerosis. Arch Neurol 56:311–315. https://doi.org/10.1001/archneur.56.3.311

Katsarou A, Gudbjörnsdottir S, Rawshani A et al (2017) Type 1 diabetes mellitus. Nat Rev Dis Primers 3:17016. https://doi.org/10.1038/nrdp.2017.16

Keenan HA, Sun JK, Levine J, Doria A, Aiello LP, Eisenbarth G, Bonner-Weir S, King GL (2010) Residual insulin production and pancreatic ss-cell turnover after 50 years of diabetes: Joslin Medalist Study. Diabetes 11:2846–2853

Kim YK, Sussel L, Davidson HW (2021) Inherent beta cell dysfunction contributes to autoimmune susceptibility. Biomol Ther 11:512. https://doi.org/10.3390/biom11040512

Kokkola A, Sjöblom SM, Haapiainen R, Sipponen P, Puolakkainen P, Järvinen H (1998) The risk of gastric carcinoma and carcinoid tumours in patients with pernicious anaemia. A prospective follow-up study. Scand J Gastroenterol 33(1):88–92. https://doi.org/10.1080/00365529850166266. PMID: 9489914

Komori A (2021) Recent updates on the management of autoimmune hepatitis. Clin Mol Hepatol 27(1):58–69. https://doi.org/10.3350/cmh.2020.0189

Konstadoulakis MM, Syrigos KN, Albanopoulos C, Mayers G, Golematis B (1994) The presence of anti-carcinoembryonic antigen (CEA) antibodies in the sera of patients with gastrointestinal malignancies. J Clin Immunol 14:310–313

Kreye J, Wenke NK, Chayka M, Leubner J, Murugan R, Maier N et al (2016) Human cerebrospinal fluid monoclonal N-methyl-D-aspartate receptor autoantibodies are sufficient for encephalitis pathogenesis. Brain 139:2641–2652. https://doi.org/10.1093/brain/aww208

Kuhn KA, Kulik L, Tomooka B et al (2006) Antibodies against citrullinated proteins enhance tissue injury in experimental autoimmune arthritis. J Clin Invest 116(4):961–973. https://doi.org/10.1172/jci25422

Lambelet M, Terra LF, Fukaya M, Meyerovich K, Labriola L, Cardozo AK, Allagnat F (2018) Dysfunctional autophagy following exposure to pro-inflammatory cytokines contributes to pancreatic beta-cell apoptosis. Cell Death Dis 9:96

Landgren AM, Landgren O, Gridley G, Dores GM, Linet MS, Morton LM (2011) Autoimmune disease and subsequent risk of developing alimentary tract cancers among 4.5 million US male veterans. Cancer 117:1163–1171

Laurent L, Clavel C, Lemaire O et al (2011) Fcgamma receptor profile of monocytes and macrophages from rheumatoid arthritis patients and their response to immune complexes formed with autoantibodies to citrullinated proteins. Ann Rheum Dis 70(6):1052–1059. https://doi.org/10.1136/ard.2010.142091

Liberal R, Zen Y, Mieli-Vergani G, Vergani D (2013) Liver transplantation and autoimmune liver diseases. Liver Transpl 19:1065–1077

Lopez-Chiriboga AS, Majed M, Fryer J, Dubey D, McKeon A, Flanagan EP et al (2018) Association of MOG-IgG serostatus with relapse after acute disseminated encephalomyelitis and proposed criterion for MOG-IgG-associated disorders. JAMA Neurol 75:1355–1363. https://doi.org/10.1001/jamaneurol.2018.1814

Lv S, Zhang J, Wu J, Zheng X, Chu Y, Xiong S (2005) Origin and anti-tumor effects of anti-dsDNA autoantibodies in cancer patients and tumor-bearing mice. Immunol Lett 99:217–227

Mack CL, Adams D, Assis DN, Kerkar N, Manns MP, Mayo MJ, Vierling JM, Alsawas M, Murad MH, Czaja AJ (2020) Diagnosis and management of autoimmune hepatitis in adults and children: 2019 practice guidance and guidelines from the American association for the study of liver diseases. Hepatology 72:671–722

Mahdi T et al (2012) Secreted frizzled-related protein 4 reduces insulin secretion and is overexpressed in type 2 diabetes. Cell Metab 16:625–633

Marignani M, Delle Fave G, Mecarocci S, Bordi C, Angeletti S, D'Ambra G, Aprile MR, Corleto VD, Monarca B, Annibale B (1999) High prevalence of atrophic body gastritis in patients with unexplained microcytic and macrocytic anemia. Am J Gastroenterol 94:766–772

Masters SL et al (2010) Activation of the NLRP3 inflammasome by islet amyloid polypeptide provides a mechanism for enhanced IL-1β in type 2 diabetes. Nat Immunol 11:897–904

Mckeon A, Pittock SJ (2011) Paraneoplastic encephalomyelopathies: pathology and mechanisms. Acta Neuropathol 122:381–400. https://doi.org/10.1007/s00401-011-0876-1

Mellemkjaer L, Andersen V, Linet MS, Gridley G, Hoover R, Olsen JH (1997) Non-Hodgkin's lymphoma and other cancers among a cohort of patients with systemic lupus erythematosus. Arthritis Rheum 40:761–768

Milanski M et al (2009) Saturated fatty acids produce an inflammatory response predominantly through the activation of TLR4 signaling in hypothalamus: implications for the pathogenesis of obesity. J Neurosci 29:359–370

Milone M (2017) Diagnosis and management of immune-mediated myopathies. Mayo Clin Proc 92:826–837. https://doi.org/10.1016/j.mayocp.2016.12.025

Moinzadeh P, Fonseca C, Hellmich M, Shah AA, Chighizola C, Denton CP, Ong VH (2014) Association of anti-RNA polymerase III autoantibodies and cancer in scleroderma. Arthritis Res Ther 16(1):R53. https://doi.org/10.1186/ar4486. PMID: 24524733; PMCID: PMC3978927

Molander S, Jønsson V, Andersen LP, Bennetzen M, Christiansen M, Hou-Jensen K, Madsen HO, Ryder LP, Permin H, Wiik A (2000) Pseudolymphoma and ventricular maltoma in patients with chronic gastritis, ulcer and Helicobacter pylori infection. Ugeskr Laeger 162:791–795

Montano-Loza AJ, Bhanji RA, Wasilenko S, Mason AL (2017) Systematic review: recurrent autoimmune liver diseases after liver transplantation. Aliment Pharmacol Ther 45:485–500

Morano A, Fanella M, Giallandro T, Di Bonaventura C (2020) Faciobrachial dystonic seizures: the borderland between epilepsy and movement disorders. Mov Disord Clin Pract 7:228–229. https://doi.org/10.1002/mdc3.12884

Mulder H, Ling C (2009) Mitochondrial dysfunction in pancreatic beta-cells in type 2 diabetes. Mol Cell Endocrinol 297:34–40

Muralidharan C, Conteh AM, Marasco MR, Crowder JJ, Kuipers J, de Boer P, Linnemann AK (2021) Pancreatic beta cell autophagy is impaired in type 1 diabetes. Diabetologia 64:865–877

von Mutius E (2007) Allergies, infections and the hygiene hypothesis–the epidemiological evidence. Immunobiology 212(6):433–439. https://doi.org/10.1016/j.imbio.2007.03.002

Nordberg LB, Lillegraven S, Lie E et al (2017) Patients with seronegative RA have more inflammatory activity compared with patients with seropositive RA in an inception cohort of

DMARD naive patients classified according to the 2010 ACR/EULAR criteria. Ann Rheum Dis 76(2):341–345. https://doi.org/10.1136/annrheumdis-2015-208873

Nyalwidhe JO, Grzesik WJ, Burch TC, Semeraro ML, Waseem T, Gerling IC, Mirmira RG, Morris MA, Nadler JL (2017) Comparative quantitative proteomic analysis of disease stratified laser captured microdissected human islets identifies proteins and pathways potentially related to type 1 diabetes. PLoS One 12:e0183908

Ohkusa T, Fujiki K, Takashimizu I, Kumagai J, Tanizawa T, Eishi Y, Yokoyama T, Watanabe M (2001) Improvement in atrophic gastritis and intestinal metaplasia in patients in whom Helicobacter pylori was eradicated. Ann Intern Med 134(5):380–386. https://doi.org/10.7326/0003-4819-134-5-200103060-00010. PMID: 11242498

Osborn O, Olefsky JM (2012) The cellular and signaling networks linking the immune system and metabolism in disease. Nat Med 18:363–374

Ospelt C, Bang H, Feist E et al (2017) Carbamylation of vimentin is inducible by smoking and represents an independent autoantigen in rheumatoid arthritis. Ann Rheum Dis 76:1176. https://doi.org/10.1136/annrheumdis-2016-210059

Pal D et al (2012) Fetuin-A acts as an endogenous ligand of TLR4 to promote lipid-induced insulin resistance. Nat Med 18:1279–1285

Patrick J, Lindstrom J (1973) Autoimmune response to acetylcholine receptor. Science 180:871–872. https://doi.org/10.1126/science.180.4088.871

Patterson CC, Karuranga S, Salpea P et al (2019) Worldwide estimates of incidence, prevalence and mortality of type 1 diabetes in children and adolescents: results from the International Diabetes Federation Diabetes Atlas, 9th edition. Diabetes Res Clin Pract 157:107842. https://doi.org/10.1016/j.diabres.2019.107842

Pescovitz MD, Greenbaum CJ, Krause-Steinrauf H et al (2009) Rituximab, B-lymphocyte depletion, and preservation of beta-cell function. N Engl J Med 361(22):2143–2152. https://doi.org/10.1056/NEJMoa0904452

Pescovitz MD, Greenbaum CJ, Bundy B et al (2014) Blymphocyte depletion with rituximab and β-cell function: twoyear results. Diabetes Care 37(2):453–459. https://doi.org/10.2337/dc13-0626

Petrie JR, Sattar N (2019) Excess cardiovascular risk in type 1 diabetes mellitus. Circulation 139(6):744–747. https://doi.org/10.1161/circulationaha.118.038137

Plebani M, Basso D (2009) Le malattie autoimmuni del tratto. Gastro-enterico. In: Tozzoli R, Bizzaro N, Villalta D, Tonutti E, Pinchera A (eds) Il Laboratorio Nelle Malattie Autoimmuni D'organo. Esculapio, Bologna, pp 313–332

Pociot F, Lernmark Å (2016) Genetic risk factors for type 1 diabetes. Lancet 387(10035):2331–2339. https://doi.org/10.1016/S0140-6736(16)30582-7

Pollack TA, Lennox BR, Muller S, Nicholson TR (2020) Autoimmune psychosis: an international consensus on an approach to the diagnosis and management of psychosis of suspected origin. Lancet Psychiatry 7:93–108. https://doi.org/10.1016/S2215-0366(19)30290-1

Querol L, Devaux J, Rojas-Garcia R, Illa I (2017) Autoantibodies in chronic inflammatory neuropathies: diagnostic and therapeutic implications. Nat Rev Neurol 2017(13):533–547. https://doi.org/10.1038/nrneurol.2017.84

Rajendran S, Anquetil F, Quesada-Masachs E et al (2020) IL-6 is present in beta and alpha cells in human pancreatic islets: expression is reduced in subjects with type 1 diabetes. Clin Immunol 211:108320. https://doi.org/10.1016/j.clim.2019.108320

Rawlings DJ, Metzier G, Wray-Dutra M, Jackson SW (2017) Altered B cell signaling in autoimmunity. Nat Rev Immunol 17:421–436. https://doi.org/10.1038/nri.2017.24

Redondo MJ, Jeffrey J, Fain PR, Eisenbarth GS, Orban T (2008) Concordance for islet autoimmunity among monozygotic twins. N Engl J Med 359(26):2849–2850. https://doi.org/10.1056/NEJMc0805398

Reindl M, Schanda K, Woodhall M, Tea F, Ramanathan S, Sagen J et al (2020) International multicenter examination of MOG antibody assays. Neurol Neuroimmunol Neuroinflamm 7:e674. https://doi.org/10.1212/NXI.0000000000000674

Richardson SJ, Willcox A, Bone AJ, Foulis AK, Morgan NG (2009) Islet-associated macrophages in type 2 diabetes. Diabetologia 52:1686–1688

Rigby MR, Harris KM, Pinckney A et al (2015) Alefacept provides sustained clinical and immunological effects in newonset type 1 diabetes patients. J Clin Invest 125(8):3285–3296. https://doi.org/10.1172/JCI81722

Rinaldi S, Davies A, Fehmi J, Wang J, Hardy TA, Barnett MH et al (2021) Overlapping central and peripheral nervous system syndromes in MOGantibody associated disorders. Neurol Neuroimmunol Neuroinflamm 8:e924. https://doi.org/10.1212/NXI.0000000000000924

Rizvi FS, Zainaldain H, Rafiemanesh H, Jamee M, Hossein-Khannazer N, Hamedifar H, Sabzevari A, Yazdani R, Abolhassani H, Aghamohammadi A et al (2020) Autoimmunity in common variable immunodeficiency: a systematic review and meta-analysis. Expert Rev Clin Immunol 16:1227–1235

Rodriguez-Calvo T (2018) Enteroviral infections as a trigger for type 1 diabetes. Curr Diab Rep 18(11):106. https://doi.org/10.1007/s11892-018-1077-2

Roep BO, Thomaidou S, van Tienhoven R, Zaldumbide A (2021) Type 1 diabetes mellitus as a disease of the β-cell (do not blame the immune system?). Nat Rev Endocrinol 17:150–161. https://doi.org/10.1038/s41574-020-00443-4

Sabater L, Gaig C, Gelpi E, Bataller L, Lewerenz J, Torres-Vega E et al (2014) A novel NREM and REM parasomnia with sleep breathing disorder associated with antibodies against IGLON5: a case series, characterization of the antigen and post- mortem studies. Lancet Neurol 13:575–586. https://doi.org/10.1016/S1474-4422(14)70051-1

Sabatino JJ, Probstel AK, Zamvil SS (2019) B cells in autoimmune and neurodegenerative central nervous system diseases. Nat Rev Neurosci 20:728–744. https://doi.org/10.1038/s41583-019-0233-2

Sabouri S, Benkahla MA, Kiosses WB et al (2020) Human herpesvirus-6 is present at higher levels in the pancreatic tissues of donors with type 1 diabetes. J Autoimmun 107:102378. https://doi.org/10.1016/j.jaut.2019.102378

von Scholten BJ, Kreiner FF, Gough SCL, von Herrath M (2021) Current and future therapies for type 1 diabetes. Diabetologia 64:1037–1048

Schulz M et al (2002) Food groups as predictors for short-term weight changes in men and women of the EPIC-Potsdam cohort. J Nutr 132:1335–1340

Scott DL, Wolfe F, Huizinga TW (2010) Rheumatoid arthritis. Lancet 376(9746):1094–1108. https://doi.org/10.1016/s0140-6736(10)60826-4

Selva-O'Callaghan A, Pinal-Fernandez I, Trallero-Araguas E, Milisenda JC, Grau-Junyent JC, Mammen AL (2018) Classification and management of adult inflammatory myopathies. Lancet Neurol 17:816–828. https://doi.org/10.1016/S1474-4422(18)30254-0

Sever D, Eldor R, Sadoun G, Amior L, Dubois D, Boitard C, Aflalo C, Melloul D (2011) Evaluation of impaired beta-cell function in nonobese-diabetic (Nod) mouse model using bioluminescence imaging. FASEB J 25:676–684

Shi J, Knevel R, Suwannalai P et al (2011) Autoantibodies recognizing carbamylated proteins are present in sera of patients with rheumatoid arthritis and predict joint damage. Proc Natl Acad Sci U S A 108(42):17372–17377. https://doi.org/10.1073/pnas.1114465108

Shi YC, Chen XJ, Zhang HM, Wang Z, Du DY (2017) Anti-N-methyl-D-aspartate receptor encephalitis during pregnancy: clinical analysis of reported cases. Taiwan J Obstet Gynaecol 56:315–319. https://doi.org/10.1016/j.tjog.2017.04.009

Shoelson SE, Lee J, Goldfine AB (2006) Inflammation and insulin resistance. J Clin Invest 116: 1793–1801

Simon GM, Cheng J, Gordon JI (2012) Quantitative assessment of the impact of the gut microbiota on lysine epsilon-acetylation of host proteins using gnotobiotic mice. Proc Natl Acad Sci U S A 109(28):11133–11138. https://doi.org/10.1073/pnas.1208669109

Sokolove J, Johnson DS, Lahey LJ et al (2014) Rheumatoid factor as a potentiator of anti-citrullinated protein antibody-mediated inflammation in rheumatoid arthritis. Arthritis Rheum 66(4):813–821. https://doi.org/10.1002/art.38307

Solans-Laqué R, Pérez-Bocanegra C, Salud-Salvia A, Fonollosa-Plá V, Rodrigo MJ, Armadans L, Simeón-Aznar CP, Vilardell-Tarres M (2004) Clinical significance of antinuclear antibodies in malignant diseases: association with rheumatic and connective tissue paraneoplastic syndromes. Lupus 13:159–164

Speliotes EK et al (2010) Association analyses of 249,796 individuals reveal 18 new loci associated with body mass index. Nat Genet 42:937–948

Stirnimann G, Ebadi M, Czaja AJ, Montano-Loza AJ (2019) Recurrent and de novo autoimmune hepatitis. Liver Transpl 25:152–166

Strickland RG, Mackay IR (1973) The reappraisal of the nature and significance of chronic atrophic gastritis. Am J Dig Dis 18:426–440

Suri JS, Danford CJ, Patwardhan V, Bonder A (2020) Mortality on the UNOS waitlist for patients with autoimmune liver disease. J Clin Med Res 9:319

Symmons D, Turner G, Webb R et al (2002) The prevalence of rheumatoid arthritis in the United Kingdom: new estimates for a new century. Rheumatology 41:793–800

Taraschenko O, Fox HS, Pittock S, Zekeridou A, Gafurova M, Eldridge E et al (2019) A mouse model of seizures in anti-N-methyl-D-aspartate receptor encephalitis. Epilepsia 60:452–463. https://doi.org/10.1111/epi.14662

Tersey SA, Nishiki Y, Templin AT, Cabrera SM, Stull ND, Colvin SC, Evans-Molina C, Rickus JL, Maier B, Mirmira RG (2012) Islet beta-cell endoplasmic reticulum stress precedes the onset of type 1 diabetes in the nonobese diabetic mouse model. Diabetes 61:818–827

Thomas NJ, Jones SE, Weedon MN, Shields BM, Oram RA, Hattersley AT (2018) Frequency and phenotype of type 1 diabetes in the first six decades of life: a cross-sectional, genetically stratified survival analysis from UK Biobank. Lancet Diabet Endocrinol 6:122–129

Toh BH (2014) Diagnosis and classification of autoimmune gastritis. Autoimmun Rev 13:459–462

Toh BH, Sentry JW, Alderuccio F (2000) The causative H+/K+ ATPase antigen in the pathogenesis of autoimmune gastritis. Immunol Today 21:348–354

Tomer Y, Shoenfeld Y (2000) Autoantibodies, autoimmunity and cancer. In: Shoenfeld Y, Gershwin ME (eds) Cancer and autoimmunity. Elsevier Science, Amsterdam, pp 141–150

Toubi E, Shoenfeld Y (2007) Protective autoimmunity in cancer (review). Oncol Rep 17(1): 245–251. PMID: 17143505

Toyka KV, Brachman DB, Pestronk A, Kao I (1975) Myasthenia gravis: passive transfer from man to mouse. Science 190:397–399. https://doi.org/10.1126/science.1179220

Trivedi PC, Bartlett JJ, Pulinilkunnil T (2020) Lysosomal biology and function: modern view of cellular debris bin. Cell 9:1131

Trouw LA, Haisma EM, Levarht EW et al (2009) Anti-cyclic citrullinated peptide antibodies from rheumatoid arthritis patients activate complement via both the classical and alternative pathways. Arthritis Rheum 60(7):1923–1931. https://doi.org/10.1002/art.24622

Tunio NA, Mansoor E, Sheriff MZ, Cooper GS, Sclair SN, Cohen SM (2021) Epidemiology of autoimmune hepatitis (AIH) in the United States between 2014 and 2019: a population-based national study. J Clin Gastroenterol 55(10):903–910. https://doi.org/10.1097/MCG.0000000000001449. PMID: 33074948; PMCID: PMC8050120

Tuomi T et al (1993) Antibodies to glutamic acid decarboxylase reveal latent autoimmune diabetes mellitus in adults with a non-insulin-dependent onset of disease. Diabetes 42:359–362

Van den Driessche A, Eenkhoorn V, van Gaal L, de Block C (2009) Type 1 diabetes and autoimmune polyglandular syndrome: a clinical review. Neth J Med 67:376–387

Vannella L, Lahner E, Osborn J, Annibale B (2013) Systematic review: gastric cancer incidence in pernicious anaemia. Aliment Pharmacol Ther 37:375–382

Velloso LA, Eizirik DL, Cnop M (2013) Type 2 diabetes mellitus—an autoimmune disease? Nat Rev Endocrinol 9:750. https://doi.org/10.1038/nrendo.2013.131

Verheul MK, van Erp SJ, van der Woude D et al (2016) Anticarbamylated protein antibodies: a specific hallmark for rheumatoid arthritis. Comparison to conditions known for enhanced carbamylation; renal failure, smoking and chronic inflammation. Ann Rheum Dis 75(8): 1575–1576. https://doi.org/10.1136/annrheumdis-2016-209248

Verpoort KN, der ZijdeCM J-v, Papendrecht-van der Voort EA et al (2006) Isotype distribution of anti-cyclic citrullinated peptide antibodies in undifferentiated arthritis and rheumatoid arthritis reflects an ongoing immune response. Arthritis Rheum 54(12):3799–3808. https://doi.org/10.1002/art.22279

Villa AR, Kraus A, Alarcon-Segovia D (2000) Autoimmune rheumatic diseases and cancer: evidence of causality? In: Shoenfeld Y, Gershwin ME (eds) Cancer and autoimmunity. Elsevier Science, Amsterdam, pp 111–117

Wang Z, Nicholls SJ, Rodriguez ER et al (2007) Protein carbamylation links inflammation, smoking, uremia and atherogenesis. Nat Med 13(10):1176–1184. https://doi.org/10.1038/nm1637

Weck MN, Brenner H (2006) Prevalence of chronic atrophic gastritis in different parts of the world. Cancer Epidemiol Biomark Prev 15:1083–1094

Weck MN, Brenner H (2008) Association of Helicobacter pylori infection with chronic atrophic gastritis: meta-analyses according to type of disease definition. Int J Cancer 123:874–881

Weis VG, Goldenring JR (2009) Current understanding of SPEM and its standing in the preneoplastic process. Gastric Cancer 12:189–197

Weissert R (2017) Adaptive immunity is the key to the understanding of autoimmune and paraneoplastic inflammatory central nervous system disorders. Front Immunol 8:336. https://doi.org/10.3389/fimmu.2017.00336

Winer S et al (2009) Normalization of obesity associated insulin resistance through immunotherapy. Nat Med 15:921–929

Winer DA et al (2011) B cells promote insulin resistance through modulation of T cells and production of pathogenic IgG antibodies. Nat Med 17:610–617

Winkler C, Raab J, Grallert H, Ziegler AG (2012) Lack of association of type 2 diabetes susceptibility genotypes and body weight on the development of islet autoimmunity and type 1 diabetes. PLoS One 7:e35410

Xiao B, Goh JY, Xiao L, Xian H, Lim KL, Liou YC (2017) Reactive oxygen species trigger Parkin/Pink1 pathway-dependent mitophagy by inducing mitochondrial recruitment of Parkin. J Biol Chem 292:16697–16708

Xiu Y, Wong CP, Bouaziz JD et al (2008) B lymphocyte depletion by CD20 monoclonal antibody prevents diabetes in nonobese diabetic mice despite isotype-specific differences in Fc gamma R effector functions. J Immunol 180(5):2863–2875. https://doi.org/10.4049/jimmunol.180.5.2863

Xu Y, Shen J, Ran Z (2020) Emerging views of mitophagy in immunity and autoimmune diseases. Autophagy 16:3–17

Yuki N, Hartung H-P (2012) Guillain-Barre syndrome. New Eng J Med 366:2294–2304. https://doi.org/10.1056/NEJMra1114525

Intestinal Flora as Initiatives of Autoimmunity

6

6.1 Mechanisms Linking Intestinal Flora with Autoimmunity

The mechanisms that relate the microbiota to autoimmune disorders are still mostly unexplored. Molecular mimicry, epitope dissemination, and bystander effects appear to be analogous to how pathogenic microorganisms alter immunological tolerance (for example, rheumatic fever caused by Streptococci) (Floreani et al. 2016; Nielsen et al. 2016). Indeed, death of intestinal epithelial cells in response to microbial infection allows self-antigens to be presented, leading to the development of autoreactive Th17 cells and other T helper cells (Campisi et al. 2016). Horai et al. (2015) have proven that commensal microorganisms stimulate retina-specific T cells in the colon, which then elicit autoimmune assault in the immune privileged ocular region, utilizing a spontaneous model of autoimmune uveitis. Furthermore, a recent study involving three infant cohorts found that children from districts with a higher prevalence of autoimmune diseases are dominated by bacterial species that produce less immunogenic lipopolysaccharide (LPS) (Vatanen et al. 2016), and that different microbiome-derived LPSs have different structures and immunogenic functions, which may influence early-life immunological education and account for the findings. It is possible that the links between the commensal microbiome and host autoimmune are complicated and multifaceted. Apparently, leveraging the gut microbiome to treat and prevent autoimmune illnesses, as well as using this knowledge to diagnose and stratify patients, still necessitates a more thorough characterization of the disease-related microbiome and a better understanding of its etiology.

Immune dysregulation in mucosal locations may have a role in the onset and progression of autoimmune disorders, according to mounting data. One theory is that when the gut barrier is compromised due to intestinal inflammation, bacterial translocation occurs, which triggers immunological responses in distant organs. Dysbiosis, on the other hand, causes immune cells to be excessively skewed, as seen in Th17 polarization. Th17 cells are most common in the intestine's lamina propria, where they release the proinflammatory cytokines interleukin (IL)-17A,

A. J. Alkhatib, *The Role of Microbes in Autoimmune Diseases*,
https://doi.org/10.1007/978-981-19-1162-0_6

IL-17F, and IL-22 to maintain gut barrier integrity and pathogen defense (Isailovic et al. 2015).

Innate immune cells are strategically distributed at the host–microbiome interface at mucosal sites, and they are the first line of defense for sensing microorganism components or products and transmitting signals into the host, eliciting responses that may alter the microbiome's composition and function (Round et al. 2010). Several mouse models of innate immune deficiency have shown dysbiosis, including mice lacking the genes MyD88 (Wen et al. 2008), Toll-like receptor 5 (TLR5) (Vijay-Kumar et al. 2010), and nucleotide-binding oligomerization domain-containing protein 2 (NBOD2) (NOD2) (Couturier-Maillard et al. 2013; Li et al. 2017). TLRs and NOD-like receptors (NLRs) are pattern recognition receptors (PRRs) that allow the host to detect conserved microbiome components. In the absence of these innate immune receptors, pathogen defense is compromised, and the tissue is predisposed to spontaneous inflammation. TLR5-deficient animals, for example, have metabolic syndrome symptoms that are linked to a changed microbiota (Vijay-Kumar et al. 2010). In mice, the lack of NOD2, an intracellular PRR for bacterial peptidoglycans, causes transmissible colitis and colitis-associated carcinogenesis, probably due to *Bacteroides vulgatus* restriction (Ramanan et al. 2014).

Rheumatoid arthritis (RA) has been studied as an example illustrating its nature as an autoimmune disease as well as the role of intestinal microbes in its occurrence. It is an immune-mediated chronic inflammatory disease. Patients usually present with synovial inflammation, which progresses to significant cartilage and bone destruction, reducing a person's ability to do simple tasks and lowering their quality of life. When compared to healthy controls, patients with RA demonstrate significant differences in the gut microbiota composition. Trillions of commensal microbacteria invade the mucosa of the human gastrointestinal tract, and they play an important role in the development, maintenance, and functioning of the host immune system. Dysbiosis of the gut microbiota can have a negative impact on the immune system both locally and throughout the host, predisposing the host to a variety of diseases, including RA. The gastrointestinal microbiota can impact the development and course of RA through proximal intestinal immunomodulatory cells, which are found in specific locations within the gut. The microbiota in the early stages of the disease seemed to be different from that of healthy controls. This discrepancy could be due to autoimmune processes. When compared to control groups, research investigations evaluating intestinal microbiota have shown that RA is associated with bacterial population growth or decline (Li and Wang 2021).

In terms of environmental variables, it has been shown that the gut microbiota has a role in the development of arthritis in mice (Horta-Baas et al. 2017). Changes in gut microbiota composition have been identified as significant actors in a variety of disease processes, the most notable of which is chronic inflammatory illnesses (Lynch and Pedersen 2016; Zechner 2017). A shift in the gut microbiota's composition, according to new research, generates a dysbiotic state that alters immune function and promotes a proinflammatory phenotype (Bartolini et al. 2020). Increased vulnerability to autoimmune disorders (such as inflammatory bowel disease, systemic inflammatory arthritis, and connective tissue pathologies), dysmetabolic syndromes, and cancers are some of the consequences (Lynch and Pedersen 2016). In humans, immunoglobulin A (IgA) anticitrullinated protein antibodies (ACPAs) have recently been shown to be detectable for several years before the clinical onset of arthritis (Nielen et al. 2004). This finding suggests that RA originates in mucosal regions like the colon and mouth. The clinical efficacy of antibacterial drugs (e.g., minocycline or salazosulfapyridine) in some RA patients supports the hypothesis that microbial populations in the gut and mouth are linked to the disease (O'Dell et al. 2001).

6.2 Intestinal Microbiota

The gastrointestinal tract houses the largest number of colonized bacteria in the human body. Several hundred species are found in everyone (Bischoff et al. 2014), with the Bacteroidetes and Firmicutes phyla accounting for more than 90% of them. However, other phyla (such as Proteobacteria, Actinobacteria, Fusobacteria, Verrucomicrobia, and Cyanobacteria) are also important in maintaining microbiota homeostasis (Mu et al. 2017). The microbiota performs a variety of functions, but one of the most important is maintaining immune system homeostasis in the host. As a result, any change in the microbiota population may have an impact on homeostasis (De Santis et al. 2015). The microbiota's adult structure is generated shortly after birth (Hu et al. 2015). Several desirable symbiotic organisms help to equilibrium in a benign and mutually beneficial manner within the bacterial community. Sensitive bacteria that are harmed by diseases, pathogenic members of the population that might cause disease, and therapeutic organisms that can help restore the status quo after a change are also present (Hu et al. 2015). Diet, probiotics, prebiotics, antibiotics, exogenous enzymes, fecal microbiota transplantation (FMT), and other environmental factors can all change the bacterial community in the gut (Kjeldsen-Kragh et al. 1991; Li and Wang 2021). The microbiota collaborates with the intestinal interface to carry out critical functions related to immunological homeostasis protection. Diet and intestinal microbiota are two critical factors that might fluctuate and alter barrier robustness and operation, with implications for intestinal permeability management. External antigens may be able to pass across the interface from the stomach cavity into the host if these materials are present (Bischoff et al. 2014). These characteristics are both a result of one's lifestyle, implying that environmental factors might influence the integrity of the intestinal interface's processes, and hence the immune response and the development of diseases like RA (Bischoff et al. 2014). The gut microbiota has been linked to the development of a variety of gastrointestinal and extraintestinal illnesses (Bragazzi et al. 2017). Furthermore, dysbiosis of gut bacteria is intimately linked to the immune system of the intestinal mucosa, which has been linked to autoimmune diseases such as RA. Variations within this bacterial population arise at an early stage of the disease

(Raczkiewicz et al. 2015), while the pathways underlying the pathophysiology of the changes remain unclear at both the cellular and molecular levels.

6.3 Intestinal Flora: Autoimmunity-Neurological Diseases

Modern sequencing methods (next generation techniques, NGS) have helped define the composition of the human gut microbiome during the last decade, allowing us to grasp the basic characteristics of a healthy gut microbiome as well as the severity of disease-related changes. Healthy gut microbiota allows the maintenance of various critical physiological activities, including the ability to regulate the innate and adaptive immune systems, according to this new understanding. Increasing evidence has linked dysbiosis to a variety of autoimmune/immune-mediated dermatological illnesses, and certain gut microbial signatures have been linked to clinical and prognostic factors in these diseases. There is evidence that specific gut microbiome profiles may play a role in regulating the clinical course of certain disorders (Colucci and Moretti 2021).

The microbiome of the human body is made up of bacteria, eukaryotes, archaea, and viruses (Lloyd-Price et al. 2016). The introduction of culture-independent approaches such as high-throughput and low-cost sequencing methods (next generation techniques, NGS) has allowed for the clarification of microbial composition in a variety of body locations (Lloyd-Price et al. 2016). The gastrointestinal tract (Thursby and Juge 2017), which contains around 1014 bacteria and is collectively referred to as the human gut microbiota (Qin et al. 2010), comprises most of the known microbes in humans.

Overall, bacteria belonging to two major phyla, Bacteroidetes and Firmicutes (Selber-Hnatiw et al. 2017), predominate in a healthy gut microbiota, which also includes eukaryotes (such as Candida, Malassezia, and Saccharomyces) (Lloyd-Price et al. 2016), Archaea (mostly belonging to the Methanobrevibacter genus) (Horz 2015), and viruses (primarily bacteriophages) (Scarpellini et al. 2015). Increased understanding of gut microbiota composition and functions in recent years has revealed that gut microbiota health is critical for achieving various physiological functions (Selber-Hnatiw et al. 2017). Food metabolism, nutrient synthesis, pathogen defense, and maintaining the integrity of the mucosal barrier are the most essential among them (Selber-Hnatiw et al. 2017). The gut microbiota's ability to modulate the innate and adaptive immune systems both locally and systemically is another feature (Lin and Zhang 2017). Recent research suggests that such regulation can also be exerted epigenetically, directly by gut microorganisms and their related metabolites, or indirectly by nutrients found in plant/animal derived diets, which can regulate the production of certain immune-modulating miRNAs (Kocic et al. 2019).

As a result, growing evidence has linked dysbiosis to a variety of illnesses, including metabolic disorders (Brial et al. 2018), cancer (Rea et al. 2018), and autoimmunity (Kosiewicz et al. 2014). In this context, a growing body of research (Salem et al. 2018; Polkowska-Pruszyńska et al. 2020) has investigated a putative involvement of gut microbiota in causing dermatological illnesses in mice and humans. Unfortunately, despite a wealth of intriguing observational evidence on the subject, a definite causative role for gut microorganisms in many disorders has yet to be discovered.

6.4 Systemic Lupus Erythematosus (SLE)

SLE is a heterogeneous autoimmune illness that affects multiple organs and has a variable clinical course (Kuhn et al. 2015). It can be diagnosed using the 2019 European League Against Rheumatism/American College of Rheumatology classification (Aringer et al. 2019), which includes both clinical and serological criteria. A malar or generalized maculopapular rash (acute LE), an annular or papulosquamous (psoriasiform) cutaneous eruption (subacute LE), and an erythematous-violaceous cutaneous lesion with secondary changes of atrophic scarring, depigmentation, and follicular hyperkeratosis/plugging (discoid LE, po) have all been identified as major clinical criterion. Available studies on the composition of the gut microbiota in SLE patients do not indicate which diagnostic criteria are met in each participant. As a result, the present data on gut microbiota in SLE is limited to patients with undefined autoimmune organ involvement of varying forms and severity. A lowered Firmicutes to Bacteroidetes ratio is the most dramatic evidence of a gut dysbiosis in SLE, according to numerous independent investigations. An examination conducted on a cohort of Spanish SLE adults (Hevia et al. 2014) discovered such data, which was later supported by other investigations conducted on Dutch (Van der Meulen et al. 2019) and Chinese patients (He et al. 2016; Wei et al. 2019).

In addition, at a lower taxonomic level, unique gut microbiome signatures have been observed in SLE patients. In this context, some researchers discovered an overall increase in gut Gram-negative bacteria in SLE (Luo et al. 2018), while others discovered an enrichment of the genera Rhodococcus, Eggerthella, Klebsiella, Prevotella, Eubacterium, Flavonifractor, and incertae sedis, as well as a depletion of the genera Dialister and Pseudobutyrivibrio. Surprisingly, such a signature proved effective in distinguishing SLE patients from controls. Despite the geographical diversity of the patients included, practically all research on the gut microbiota in SLE show specific changes in the Firmicutes to Bacteroidetes ratio. Such data could point to the gut microbiota of people with SLE being independent of their diet or behaviors. In SLE, certain bacterial enrichments have also been identified as indicators of activity. As a result, increased levels of Streptococcus Campylobacter and Veillonella have been linked to disease activity, while Bifidobacterium has been linked to remission (Li et al. 2019). Furthermore, the SLE disease activity index (SLEDAI) (Azzouz et al. 2019), a validated measure of SLE activity (Bombardier et al. 1992), has been shown to be strongly correlated with a decrease in total gut bacterial richness. The above-mentioned dysbiosis and microbial enrichment/depletion could be to blame for SLE patients' decreased synthesis of short chain fatty acids (SCFAs), primarily acetate and propionate (Rodríguez-Carrio et al. 2017). SCFAs are involved in gut barrier modification (Morrison and Preston 2016) and inflammatory/immune response regulation and are produced by bacteria under physiological settings because of their metabolic and fermentative processes. The abnormal release of SCFAs reported in SLE patients suggests possible deficiencies in gut bacteria-mediated immune system modulation, which could contribute to the aberrant autoimmune activation seen in SLE pathogenesis. In addition to indirect activation of important immune-pathogenic pathways in SLE, certain gut bacteria appear to play a direct activating role. A study demonstrated that adding feces from SLE patients to naive CD4? cells could enhance lymphocyte activation and polarization toward the Th17 phenotype (López et al. 2016), which is a well-known T-cell fraction that plays a key role in SLE pathogenesis. Furthermore, in SLE patients, a direct link was discovered between fecal Firmicutes abundance and serum Th1 cells and IFN-c, which was not seen in controls (López et al. 2016). A study on intestinal *Ruminococcus gnavus* found a fivefold increase in gut bacteria in SLE patients compared to controls (Azzouz et al. 2019), which could indicate a direct bacterial pathogenic function. SLE patients exhibit circulating antibodies directed against B-cell superantigens of the cell wall lipoglycans of a specific strain of this bacterium, in addition to gut enrichment. Antibodies against native DNA titers were found to be directly associated to disease activity, active lupus nephritis, and anti-native DNA titers (Azzouz et al. 2019). Such evidence implies that R. gnavus gut enrichment in SLE could cause B-cell activation and, as a result, pro-inflammatory consequences (Bunker et al. 2019; Silverman et al. 2019). It's possible that a compromised gut barrier (leaky gut syndrome) could promote the translocation beyond the gastrointestinal barrier of certain immunogenic bacteria strains with epitopes that can cross-react with specific self-antigens via molecular mimicry processes, triggering systemic autoimmune pathways (Kim et al. 2019). Interestingly, SLE patients have been found to have high amounts of fecal calprotectin (Azzouz et al. 2019), a sign of intestinal inflammation and mucosal injury. Furthermore, an in vitro study by Greiling et al. (2018) found that sera from SLE patients positive for autoantibodies directed toward the RNA binding protein Ro60 (anti-nuclear antibodies antiRo60) also undergo T- and B-cell activation as well as the release of pro-inflammatory cytokines after in vitro stimulation by selected cAMP. Intestinal Bacteroides thetaiotaomicron, a bacterium with a high sequence similarity to human Ro60 protein (Ro60 ortholog) (Greiling et al. 2018) was identified as one of these reactive bacteria. A cross-reactivity phenomenon between bacterial and human epitopes has been discovered according to this study. The dysbiosis and specific gut microbiome profiles discovered so far could help explain why women are more likely to develop SLE (Christou et al. 2019; Krasselt and Baerwald 2019). In mice, it has been shown that gender has a significant impact on particular gut microbial genetic and metabolic pathways involved in immune control, with estrogens boosting the immune system and androgens suppressing it (Markle et al. 2013; Johnsona et al. 2020). In this context, certain gut microbiome profiles may be linked to the highest hyper-reactive immune responses, resulting in greater sensitivity to autoimmunity in females compared to males (Vemuri et al. 2019). Only mice have been used to study the differences in gut microbiota composition in females and males with SLE. Female mice, as expected, had a different gut microbiota composition than male rodents, which was linked to a faster illness progression in females (Vemuri et al. 2019).

6.5 Atopic Dermatitis (AD)

AD is a chronic inflammatory skin disorder that primarily affects children and is characterized by persistent itching and age-specific skin manifestations (Kapur et al. 2018). AD can sometimes extend into maturity or even start in adulthood (Megna et al. 2017). Increased IgE levels have been linked to Alzheimer's disease (Yang et al. 2020). This measure distinguishes between intrinsic (normal IgE, non-allergic) and extrinsic (high IgE, allergic, and more severe) AD (Yang et al. 2020). Extrinsic AD patients appear to have a higher risk of having the so-called atopic march, a well-defined progression of disorders that begins with atopic dermatitis and food allergies in childhood and progresses to allergic asthma and allergic rhinitis later in life (Yang et al. 2020). AD derives from an intricate interaction between genes and environment. Pathogenetically, abnormalities of intercellular lipids, filaggrin, and tight junctions cause a breaking of the skin barrier, ultimately developing into skin inflammation (Nakahara et al. 2020). Even if it is widely recognized that Th2 cells and their related cytokines play a major role in AD inflammation, mostly in the acute phase, recent findings suggest an adjunctive switch of the T response toward a type 1/Th17 phenotype, especially in the chronic phase of AD (Gittler et al. 2012; Clayton et al. 2020). The hygiene hypothesis (Kim et al. 2019) was one of the first to identify a possible link between microorganisms and AD. Given the rising prevalence of allergy disorders in modern Westernized populations, this theory proposes that inadequate microbial exposure early in life may result in poorer immunological priming and, as a result, a higher chance of developing allergic or autoimmune disorders later in life. In this situation, a eubiotic gut microbiota throughout early childhood might provide proper immune tolerance and prevent allergic over-sensitizations due to the known ability of gut bacteria to influence immune responses toward pathogens and tolerance (Penders et al. 2007b). Following that, studies using quantitative real-time PCR and/or denaturing gradient gel electrophoresis (DGGE) revealed differences in the types and abundance of bacteria inhabiting the gut of AD patients compared to healthy controls, implying that gut microbiota composition may play a role in AD development (Penders et al. 2007a, b). Metagenomics has unquestionably produced clear and comprehensive evidence of a changed gut microbiota composition in Alzheimer's patients. In AD-prone children, gut dysbiosis appears to be an early and long-lasting event that might induce immunological activation and cytokine release, which can lead to the development of AD clinical symptoms. An NGS investigation on pregnant atopic women and their offspring by West et al. (2015) demonstrated such an occurrence. The authors looked at the gut microbiota of such infants at baseline (at 1 week, 1 month, and 1 year) and compared it to specific markers of innate immune responses at 6 months (measured as cytokine production from peripheral blood mononuclear cells after activation with specific microbial ligands for TLR2

(Pansorbin) and TLR4 (lipopolysaccaride) and the development of IgE-associated eczema. Atopic patients had dysbiosis and immunological activation, as evidenced by a lower relative abundance of Ruminococcaceae at 1 week in children developing future IgE-associated dermatitis and higher IL-6 and TNF-a production after TLR2-ligand stimulation (West et al. 2015). TLR4-induced TNFa secretion was also observed to have an inverse relationship with Proteobacteria abundance (at 1 week and 1 month) and Enterobacteriaceae enrichment (at 1 year) (West et al. 2015). A subsequent analysis revealed that AD children have a long-term gut bacterial impairment, characterized by a reduction in microbial diversity (Abrahamsson et al. 2012). Early in life, some variables can influence the composition of the gut microbiota. The route of birth (cesarean versus vaginal) (Dominguez-Bello et al. 2010), the kind of feeding (Ho et al. 2018), the composition of the maternal gut microbiota during pregnancy (Hong et al. 2010), and the existence of an older sibling (Madsen et al. 2010; Laursen et al. 2015) are among them. In children, vaginal delivery and nursing have been found to be protective against the development of AD, while expectant mothers whose infants later developed atopy have been found to have a lower diversity and quantity of gut flora (Hong et al. 2010; West et al. 2015). The presence of an older brother appears to have no effect on the risk of Alzheimer's disease in youngsters (Laursen et al. 2015). All the above-mentioned possible confounding factors should be considered in the enrolling process for studies investigating gut microbiota in Alzheimer's disease. Bacteria have also been discovered as potential sustainers of a gut dysbiosis in Alzheimer's disease. Faecalibacterium and Ruminococcus gnavus (Zheng et al. 2016; Chua et al. 2018; Reddel et al. 2019), whose propensity to induce pro-inflammatory and immuno-sensitizing responses, as established in mice (Chua et al. 2018), suggesting a potential pathogenic role, are the bacteria most identified as abundant in AD children by numerous independent studies. In AD youngsters, a reduction of Bifidobacterium, a bacterium capable of generating anti-inflammatory SCFAs, has been often seen (Zheng et al. 2016; Reddel et al. 2019). Supplementary bacteria have been reported to be increased in children with Alzheimer's disease (Zheng et al. 2016; Reddel et al. 2019), while others have been found to be decreased (Lee et al. 2018a, b). Among them, reduction of *Akkermansia muciniphila* in AD children has been linked to altered expression of functional genes involved in immune system control (Lee et al. 2018a, b). These genes are involved in the phosphoinositide 3-kinase-protein kinase B (PI3K-Akt) signaling pathway, which has been linked to epithelial cell and dendritic cell survival, as well as the nucleotide-binding oligomerization domain (NOD)-like receptors signaling pathway, which is linked to gut microbiota homeostasis. As a result of such gene regulation, abnormal antigen processing and presentation may occur, encouraging immunological sensitization. Furthermore, because *Akkermansia muciniphila* is involved in mucus degradation, its depletion in AD may promote an increased mucus layer thickness, which reduces the availability of nutrients (such as glycans) for other potentially beneficial microbes, resulting in bacterial dysbiosis (Lee et al. 2018a, b). So far, no changes in gut microbiota composition have been discovered in AD children from rural Africa (Mahdavinia et al. 2017). Indeed, when compared to non-atopic African children, the unique study conducted on such atopic newborns found no variations in bacterial richness and abundance of species. According to the authors, AD in African youngsters is more likely to be caused by genetically predisposed backgrounds or skin barrier deficits than by gut dysbiosis. To fully understand such a link, more in-depth studies are required. Except for the above-mentioned study, the findings of early alterations in gut microbiota composition or enrichment/depletion of bacteria involved in mucus layer homeostasis in AD studies so far are consistent with the findings of early alterations in gut microbiota composition or enrichment/depletion of bacteria involved in mucus layer homeostasis. Dysbiosis, gut inflammation, and increased permeability may result because of these changes, allowing pro-sensitizing toxins, bacteria, or antigens to enter the bloodstream and stimulate immune system sensitization processes, potentially contributing to the development of AD (Lee et al. 2018a, b; Petersen et al. 2019). Furthermore, certain bacteria may release circulating neurotransmitters capable of influencing local itch and inflammatory processes (Lee et al. 2018a, b).

Most studies have focused on the gut microbiota composition in AD children so far, because this disease mostly affects children under the age of puberty and usually vanishes in adults (Kapur et al. 2018). However, an increase in the incidence of adult Alzheimer's disease has lately been documented (Sacotte and Silverberg 2018).

6.6 Psoriasis (PS)

Ps is a systemic and chronic inflammatory skin disease characterized by erythematous plaques protected by silvery scales that appear in pathognomonic body regions. In the etiology of the disease, the immune system, environmental factors, and genetic background all play a part. From a molecular basis, Ps is defined by the invasion of the epidermis by activated T lymphocytes capable of stimulating keratinocyte development (Nair and Badri 2020). Such an occurrence is caused by a complex, and in part unknown, synergism of pathways comprising inflammation, cell signaling, antigen presentation, and transcriptional control (Grän et al. 2020). The study of the gut microbiota composition in Ps has only recently begun, and most of the studies have focused on the most common variety of Ps, Ps vulgaris. The gut microbiota of Ps appears to be dysbiotic, as it has been described as having a lower microbial diversity (Hidalgo-Cantabrana et al. 2019; Shapiro et al. 2019) and a lower Bacteroidetes to Firmicutes ratio (Huang et al. 2019). Different independent studies using comparable methods have identified specific gut microbiota signatures characterized by a depletion of Coprococcus (Scher et al. 2015) and *Akkermansia muciniphila* (Scher et al. 2015) and an enrichment of Faecalibacterium (Chen et al. 2018), Ruminococcus (Codoñer et al. 2018), Megasphaera (Codoñer et al. 2018), Actinobacteria (Shapiro et al. 2019), *Dorea formicigenerans* (Shapiro et al. 2019), and *Colinsella aerofaciens* (Shapiro et al. 2019). Some of the abovementioned gut microbiota in Ps elicit certain metabolic pathways with pro-inflammatory effects from a functional standpoint. Coprococcus reduction is associated with a decrease in heptanoate and hexanoate in the feces, both of which are beneficial SCFAs, as well as an increase in the proinflammatory protein RANKL and secretory IgA, both of which are markers of intestinal and systemic inflammation (Scher et al. 2015). In the same study, comparable changes were discovered in a subgroup of psoriatic patients who developed arthritis (Scher et al. 2015), a rheumatological illness that affects 30% of Ps patients (Ritchlin et al. 2017). The beneficial bacteria Pseudobutyrivibrio, Ruminococcus, and *Akkermansia muciniphila* are also depleted in psoriatic arthritis (Scher et al. 2015). Another investigation on Ps patients without articular involvement revealed the latter, though at a lesser abundance (Tan et al. 2018).

In contrast, several investigations found no significant differences in gut microbiota composition between psoriatic arthritis and Ps vulgaris (Huang et al. 2019). Firmicutes and Actinobacteria (phyla) enrichment, as well as *Dorea formicigenerans* and *Colinsella aerofaciens* (species) enrichment (Shapiro et al. 2019), coincides with a notable increase in metabolic pathways involved in LPS function. LPS has been linked to intestinal inflammation, and its increased levels have been linked to insulin resistance and diabetes mellitus, both of which are common Ps comorbidities (Shapiro et al. 2019).

In psoriatic patients, gut microbiota studies revealed signs of mucus layer and intestinal barrier damage. The decreased abundance of intestinal *Akkermansia muciniphila* (Tan et al. 2018) stands out among them. This bacterium is important for gut intestinal barrier homeostasis and eubiosis, as it influences the thickness and glycosylation pattern of the mucus layer, which influences the quantity and types of resident microbes (Ouwerkerk et al. 2013). Specifically, an increase in plasma I-FABP correlates with the well-known PASI (Sikora et al. 2019) indicator of Ps severity, whereas an increase in Veillonella coincides with an increase in blood C-reactive protein (Huang et al. 2019), a marker of systemic inflammation that is commonly elevated in Ps (Beygi et al. 2014). A leaky gut syndrome caused by increased gut permeability could also lead to bacterial translocation from the gut to the circulation (Ramírez-Boscá et al. 2015; Visser et al. 2019). Psoriatic patients have been found to have bacterial DNA fragments, primarily belonging to E. coli. They have been linked to systemic inflammation, which is defined by elevated circulating levels of IL-1b, IL-6, IL-12, tumor necrosis factor, and interferon c, as well as a more severe psoriasis course (Ramírez-Boscá et al. 2015). This research establishes a bacterial trigger for a systemic inflammatory response and a change in clinical course in Ps (Ramírez-Boscá et al. 2015). Increased permeability indicators and particular microbial signatures, taken together, give proof of a general state of microbiota-driven inflammation in Ps, implying that gut microbiota may play a pathogenic role in Ps etiology.

6.7 Alopecia Areata (AA)

Alopecia areata (AA) is a kind of alopecia that causes non-cicatricial hair loss on the scalp, beard, body, eyebrows, and eyelashes because of an autoimmune attack on hair follicles. Its pathophysiology involves local activation of the Th1 and Th17 pathways, which results in the release of pro-inflammatory cytokines and peribulbar inflammation, which leads to hair loss (Lepe and Zito 2020). The anecdotal observations of Rebello et al. (2017) on patients undergoing fecal transplantation for gastrointestinal disease treatment provided the first indication of a potential involvement of gut microbiota makeup in modifying the immunological pathways involved in AA progression. The authors described two cases of young adults with active AA who were resistant to standard therapy and underwent fecal transplantation for simultaneous Clostridium difficile colitis and Crohn's disease (Rebello et al. 2017). Both patients exhibited moderate hair regrowth, which lasted after a long follow-up, in addition to the expected satisfactory result of their gastrointestinal illnesses. Other researchers recently reported a similar finding in an older patient who had complete and long-lasting hair growth after fecal transplantation for noninfectious diarrhea (Xie et al. 2019). Only one study has used next-generation sequencing technologies to investigate gut microbiota composition in AA thus far (Moreno-Arrones et al. 2020). The researchers conducted a cross-sectional investigation with 15 patients who had AA universalis, a more severe form of AA that causes full hair loss on the scalp and body. The relative abundance of species in the studied groups revealed an interesting enrichment of Erysipelotrichaceae, which can trigger the release of proinflammatory cytokines and Lachnospiraceae, which have previously been linked to some autoimmune AA comorbidities (AD, sclerosing cholangitis, and anchylosing spondylitis) ((Moreno-Arrones et al. 2020). Furthermore, a decrease in the number of bacteria of the Clostridiales order that make SFCAs was identified, implying a possible loss of such bacteria's antiinflammatory powers (Moreno-Arrones et al. 2020). The decision to include only AA universalis patients is notable because the discovery of the above-mentioned gut bacterial signature in other, less severe forms of AA could indicate an adverse progression toward a universalis type.

6.8 Microbial Therapeutics for Halting and Prevention of Autoimmune Disease

Regulatory T cells (Treg) are important for maintaining peripheral tolerance as well as controlling chronic inflammation and autoimmune. The cytokine interleukin-2 (IL-2) is needed for Treg survival and proliferation in peripheral lymphatic tissues and so plays an important role in Treg biology. Treg biology is disrupted in most autoimmune and rheumatic illnesses, either numerically or functionally, resulting in an imbalance between protective and harmful immune cells. Furthermore, during the pathogenesis of several autoimmune disorders, a relative deficit of IL-2 develops, disrupting Treg homeostasis and amplifying the vicious cycle of tolerance breach and chronic inflammation. Low-dose IL-2 therapy attempts to either compensate for the IL-2 shortage and restore a physiological condition, or to boost the Treg population so that it can be more successful in counter-regulating inflammation while avoiding global immunosuppression (Graßhoff et al. 2021).

Since the discovery and cloning of the cytokine interleukin-2 (IL 2) in the 1970s, the understanding of its function and role in the immune system has shifted dramatically (Graßhoff et al. 2021). Because of its ability to increase T cell proliferation in vitro, IL-2 was once thought to be a critical element in the development of inflammatory immune responses against invading pathogens and tumors and was thus used to treat malignant disorders at large doses (Malek 2008). A key finding that cast doubt on the early concept of IL-2 was that mouse genetically lacking for IL-2 or IL-2 receptor components developed broad and deadly autoimmune disorders due to uncontrolled T and B cell hyperactivity (Malek and Castro 2010; Grashoff et al. 2010), rather than the expected immunological deficiency. Later studies were able to link IL-2 to immune tolerance by demonstrating that it is essential for the growth and survival of regulatory T cells (Treg) in the peripheral lymphatic organs, as well as their thymic development and differentiation (Bayer et al. 2007), highlighting the critical role of IL-2 in Treg biology. As a result, IL-2 is now more commonly referred to as a "immune regulatory" cytokine, and it may play a much less role in the development of pro-inflammatory and anti-tumor immune responses than previously thought (Malek and Castro 2010). This novel and scientifically supported viewpoint paved the way for therapeutic exploration of the Treg-IL-2 axis in the context of immune-mediated and inflammatory diseases, with the goal of directly increasing the Treg population in the patient, thereby counteracting pathogenic autoimmune responses and re-establishing immune tolerance. In further investigations, IL-2 was found to suppress germinal center formation and autoantibody synthesis by reducing the development of T follicular helper (Tfh) cells independent of Treg interference (Ballesteros-Tato et al. 2012; Ballesteros-Tato 2014), as well as to limit the differentiation of naive helper T cells into Th17 cells (Laurence et al. 2007). In a wide range of autoimmune and rheumatic disorders, these CD4+ T cell subsets are thought to play a pathogenic role. Because of the long-term approval of large doses of IL-2 for cancer therapy, the notion of employing modest doses of IL-2 for the treatment of immunological illnesses was developed. Treg are more IL-2 sensitive and require considerably lower doses of IL-2 in comparison to anti-tumor T cells and NK cells, they were stimulated by IL-2. Because they express high levels of the protein on a constant basis, cells expressing IL-2 receptor make complex that is heterotrimeric and has a high affinity CD25 (a-chain), CD122 (b-chain), and CD132 (c-chain), (G-chain) (common g-chain) (Sakaguchi 2005; Klatzmann and Abbas 2015).

6.9 Rationales for IL-2 Therapy in Autoimmune and Rheumatic Diseases

Tregs that express the lineage-specific transcription factor FoxP3 are critical for maintaining immunological selftolerance and, as a result, for preventing and controlling autoimmune disorders (Miyara et al. 2011). FoxP3+ Treg are primarily produced from a separate CD4+ T cell subset in the thymus and are essential to regulate the activation, differentiation, and expansion of autoreactive T cells and other potentially damaging immune cells in the peripheral lymphatic organs (Sakaguchi et al. 2010). As a result, it is reasonable to suppose that a disruption of Treg biology, either numeric or functional, plays a role in the pathogenesis of most rheumatic and autoimmune illnesses (Sharabi et al. 2018). A relative shortfall or shortage of IL-2 can occur in autoimmune disorders, leading to a modification of Treg homeostasis, which further accelerates the vicious cycle of tolerance breach and chronic inflammation (von Spee-Mayer et al. 2016). Low-dose IL-2 therapy tries to correct for this IL-2 deficiency or to enhance the Treg population so that it can be more successful in regulating inflammation while avoiding general immunosuppression (Humrich and Riemekasten 2016a, b). Systemic lupus erythematosus (SLE) appears to be an appropriate and prospective candidate disease for a therapeutic intervention with low-dose IL-2 therapy from an immune-pathophysiological standpoint. SLE is a type of systemic autoimmune illness with an unknown cause that is defined by a rupture of tolerance to a wide range of nuclear autoantigens, resulting in inflammation in various organs (Tsokos 2020).

Taken together, low-dose IL-2 therapy is very safe and capable of selecting expanding a functionally competent Treg population independent of the underlying disease, according to data from multiple pilot investigations and clinical trials, including the first randomized trials. Furthermore, these studies revealed preliminary evidence of low-dose IL-2 therapy's clinical usefulness in a wide range of inflammatory and autoimmune illnesses. Low-dose IL-2 therapy might thus be thought of as a novel targeted treatment approach with a wide range of potential applications in autoimmune, inflammatory, and rheumatic illnesses. Variations in clinical responsiveness between diseases or patient subgroups could be related to differences in the form of Treg abnormalities or the amount to which they contribute to disease etiology. Disease-related alterations in IL-2 signaling pathways and associated molecules that lead to differences in biological responsiveness to IL-2 therapy could potentially affect clinical efficacy, but this is unlikely due to the relatively universal response pattern to low-dose IL-2 therapy reported thus far. Despite some clinical heterogeneity, which stems from significant differences in study design, treatment regimens, and treatment duration, the results of most of these trials justify further investigation of this novel therapeutic approach in autoimmune and rheumatic diseases and provide a valuable scientific foundation for placebo-controlled and larger confirmatory trials. Advanced immunophenotyping technologies will allow for the identification of molecular, cellular, and epigenetic key events in response to low-dose IL-2 therapy at a common and disease-specific level, as well as biomarkers that can predict biological and clinical responsiveness to low-dose IL-2 therapy, allowing for the selection of appropriate diseases or patient subgroups and stratification of patients based on their clinical status (74). Modified IL-2 formulations with a longer half-life or higher selectivity for Treg in the clinic could help maintain clinical and biological efficacy, including Treg lineage and function stability, while also making it more accessible to patients. Aside from that, low-dose IL-2 therapy appears to be an ideal candidate for a combination therapy, such as with agents that block the activity of inflammatory cytokines and pathways, which can also promote Tcon resistance to Treg-mediated suppression, or with B cell directed therapies, due to its excellent safety profile and unique mode of action (Graßhoff et al. 2021).

6.10 The Crosstalk Between Intestinal Microbiome and Host and Adaptive Immunity

IgA is the most prevalent secretory Ig isotype in the gut, and it can be generated without or with the help of T cells (Fagarasan et al. 2010). Segmented filamentous bacterium (SFB) and Alcali genes are notable examples of commensal microbial stimulation required for intestinal IgA responses (Lecuyer et al. 2014). The intestinal microbiota of low-IgA animals, on the other hand, has been found to degrade both the secretory component of secretory IgA and IgA itself, which could explain their colitis susceptibility (Moon et al. 2015). IgA plays a pivotal role in mucosal defense by covering and entrapping bacteria and, additionally, immobilizing the microbiome by downregulating flagella-related gene expression (Cullender et al. 2013).

According to modern approaches in sorting and sequencing immunoreactive pathosymbionts, IBD may be initiated preferentially by commensals with high IgA coating, an intriguing working hypothesis (termed IgA-seq) (Palm et al. 2014). The colonization of an IgA coated Escherichia coli, which could generate Th17 mucosal immunity, was observed to be preferentially enriched in individuals with Crohn's disease and associated spondyloarthritis (Viladomiu et al. 2017). IgA coating facilitates non-invasive microbial translocation by promoting antigen presentation and subsequent synthesis of antigen-specific IgA (Fransen et al. 2015). These IgA molecules, which have gone through somatic hypermutation and affinity maturation, bind to and select certain bacteria. As a result, IgA plays a role in shaping and maintaining the microbial community. Other immune cells that influence gut microbial diversity through IgA selection in Peyer patches include Treg cells and follicular helper T (Tfh) cells (Kawamoto et al. 2014). Adaptive immunity is partially impaired in germ-free (GF) and antibiotic-treated mice, as evidenced by a lack of intestinal Th17 and Treg cells and a skewing toward Th2. Monocolonization of mice with SFB is sufficient to generate intestinal Th17 cells, rheumatoid arthritis (RA), and multiple sclerosis (MS) (Gaboriau-Routhiau et al. 2009; Wu et al. 2010; Lee et al. 2011). Furthermore, antinuclear antibody generation and subsequent systemic autoimmunity in adult mice have been demonstrated to be influenced by neonatal SFB colonization and related IL-17 signaling (Van Praet et al. 2015). Following examinations into SFB, it was discovered that Th17 cell responses to the antigen are at least largely antigen specific (Yang et al. 2014). MHC-II-dependent presentation of SFB antigens by conventional DCs has been shown to induce mucosal Th17 cell development (Goto et al. 2014). SFB's direct attachment to the ileal epithelium also causes the production of serum amyloid A proteins 1 and 2 (SAA1/2).

SFB also stimulates ILC3 to generate IL-22, which boosts epithelial SAA production even more (Sano et al. 2015). However, it is unclear how SAA affects Th17 cells. CX3CR1+ myeloid cells may respond to SAA by secreting cytokines that stimulate the polarization of Th17 cells and the generation of IL-22 by ILC3 (Panea et al. 2015). Even though SFB has yet to be discovered in human intestines, *Bifidobacterium adolescentis*, a human-derived symbiotic bacterial species, serves as an equivalent strong inducer of Th17 cells in mouse intestines via a non-SFB mechanism (Tan et al. 2016).

High salt consumption, on the other hand, can cause Lactobacillus murinus depletion, which can lead to Th17 dysregulation and autoimmune diseases, which can be treated with L. murinus supplementation (Wilck et al. 2017). A pilot human study found the same drop in Lactobacillus spp. and an increase in Th17 cells because of increased salt intake, indicating a link between food and the gut–immune axis. To resist pathogenic infections while maintaining tolerance to commensals, mucosal immunity necessitates a careful balance. Treg cells maintain intestinal homeostasis by preventing aberrant immune responses to food antigens and the commensal microbiome, therefore preventing the onset of immunopathology. Certain elements of the commensal microbiome can produce and retain intestinal Treg cells, as evidenced by the fact that the Bacteroides fragilis-derived polysaccharide, PSA, can repair immunologic deficit in GF mice (Tanoue et al. 2016). PSA is a symbiotic factor that can convert CD4 + T cells into IL-10-producing Tregs and reduce mucosal inflammation in mice (Telesford et al. 2015). This immunomodulatory impact is dependent on the activation of a non-canonical autophagy pathway by two IBD-associated genes, ATG16L1 and NOD2, which could explain the lack of Treg responses in people with these risk genes (Chu et al. 2016). PSA has also been shown to have anti-inflammatory properties in extraintestinal autoimmune mice models, such as multiple sclerosis (Wang et al. 2014).

Atarashi et al. (2013) showed that a specific group of Clostridia strains can induce Treg cells in the mouse gut, supporting the hypothesis that local microbial populations are substantially responsible for the tolerogenic cell type. SCFAs, particularly butyrate, may be involved in the microbial activation of Tregs via HDAC inhibition and subsequent histone H3 acetylation of the Foxp3 gene (Furusawa et al. 2013). SCFAs, on the other hand, boost Treg cell proliferation by activating G-protein-coupled receptors like GPR43 (Smith et al. 2013). Recent research has added to the intricacy of Tregs and Th17 cells produced by symbionts. Two research groups have discovered a subpopulation of Tregs that lack NRP1 and Helios but, curiously, express RORt, a transcription factor that is thought to counteract FoxP3 and promote Th17 cell development (Ohnmacht et al. 2015). In experimental colitis, this FoxP3+ RORt+ Treg subtype expresses high amounts of IL-10 and CTLA-4 and demonstrates improved suppressive ability (Yang et al. 2016). In the intestine, another population of GATA3-driven Treg cells is generated by epithelium-derived IL-33 (Schiering et al. 2014). The balance between GATA3-expressing and RORt-expressing Treg cells is regulated by the microbiota and other tissue-derived stimuli, but the latter cells are not localized. Changing the Treg balance in the intestine because of numerous environmental stimuli could have systemic consequences, prolonging autoimmune damage and chronic inflammation.

Intriguingly, the microbiome has been hypothesized to play a key role in lymphopenia-associated autoimmune, which could explain the contradictory coexistence of autoimmunity and immunodeficiency within a single person. During lymphopenia, peripheral T cells engage in a process known as "homeostatic proliferation," which helps to sustain the immune system while also increasing the risk of autoreactive clones expanding abnormally (Jones et al. 2013). In this situation, a two-hit model based on commensal microbiota has been proposed to explain lymphopenia-associated colitis etiology (Feng et al. 2010). First, the microbiota increases the synthesis of IL-6 by innate cells via MyD88, which provides signals for T cells to proliferate spontaneously.

These T cells proliferate in an antigen-specific way in the presence of microbiota, resulting in colon inflammation. In MyD88- and TLR-deficient mice models, however, T cell homeostatic proliferation was found to be inconsistently normal or even enhanced (Cording et al. 2013).

Eri et al. (2017) have demonstrated that T cells transplanted into athymic mice grow and develop into a unique PD-1 + CXCR5 /dim T cell fraction, which acts as Tfh cells to drive B cell autoantibody synthesis. Antibiotic depletion of the commensal microbiota inhibits differentiation and improves systemic autoimmunity (Eri et al. 2017).

References

Abrahamsson TR, Jakobsson HE, Andersson AF, Bjorksten B, Engstrand L, Jenmalm MC (2012) Low diversity of the gut microbiota in infants with atopic eczema. J Allergy Clin Immunol 129: 434–440

Aringer M, Costenbader K, Daikh D, Brinks R, Mosca M, Rmasey-Goldman R et al (2019) European league against rheumatism/American college of rheumatology classification criteria for systemic lupus erythematosus. Arthritis Rheumatol 71:1400–1412

Atarashi K, Tanoue T, Oshima K, Suda W, Nagano Y, Nishikawa H et al (2013) Treg induction by a rationally selected mixture of clostridia strains from the human microbiota. Nature 500:232–236

Azzouz D, Omarbekova A, Heguy A, Schwudke D, Gisch N, Rovin BH et al (2019) Lupus nephritis is linked to disease-activity associated expansions and immunity to a gut commensal. Ann Rheum Dis 78:947–956

Ballesteros-Tato A (2014) Beyond regulatory T cells: the potential role for IL-2 to deplete T-follicular helper cells and treat autoimmune diseases. Immunotherapy 6:1207–1220. https://doi.org/10.2217/imt.14.83

Ballesteros-Tato A, Leon B, Graf BA, Moquin A, Adams PS, Lund FE et al (2012) Interleukin-2 inhibits germinal center formation by limiting T follicular helper cell differentiation. Immunity 36:847–856. https://doi.org/10.1016/j.immuni.2012.02.012

Bartolini I, Risaliti M, Ringressi MN, Melli F, Nannini G, Amedei A, Muiesan P, Taddei A (2020) Role of gut microbiota-immunity axis in patients undergoing surgery for colorectal cancer: focus on short and long-term outcomes. World J Gastroenterol 26(20):2498–2513. PMID: 32523307; PMCID: PMC7265137. https://doi.org/10.3748/wjg.v26.i20.2498

Bayer AL, Yu A, Malek TR (2007) Function of the IL-2R for thymic and peripheral CD4+CD25+ Foxp3+ T regulatory cells. J Immunol 178:4062–4071. https://doi.org/10.4049/jimmunol.178.7.4062

Beygi S, Lajevardi V, Abedini R (2014) C-reactive protein in psoriasis: a review of the literature. J Eur Acad Dermatol Venereol 28:700–711

Bischoff SC, Barbara G, Buurman W, Ockhuizen T, Schulzke JD, Serino M, Tilg H, Watson A, Wells JM (2014) Intestinal permeability—a new target for disease prevention and therapy. BMC Gastroenterol 14:189. https://doi.org/10.1186/s12876-014-0189-7S

Bombardier C, Gladman DD, Urowitz MB, Caron D, Chang CH (1992) Derivation of the SLEDAI: a disease activity index for lupus patients. Arthritis Rheum 35:630–640

Bragazzi NL, Watad A, Neumann SG, Simon M, Brown SB, Abu Much A, Harari A, Tiosano S, Amital H, Shoenfeld Y (2017) Vitamin D and rheumatoid arthritis: an ongoing mystery. Curr Opin Rheumatol 29(4):378–388. PMID: 28463872. https://doi.org/10.1097/BOR.0000000000000397

Brial F, Le Lay A, Dumas ME, Gauguier D (2018) Implication of gut microbiota metabolites in cardiovascular and metabolic diseases. Cell Mol Life Sci 75:3977–3990

Bunker JJ, Drees C, Watson AR, Plunkett CH, Nagler CR, Schneewind O et al (2019) B cell superantigens in the human intestinal microbiota. Sci Transl Med 11:9356. https://doi.org/10.1126/eaau9356

Campisi L, Barbet G, Ding Y, Esplugues E, Flavell RA, Blander JM (2016) Apoptosis in response to microbial infection induces autoreactive TH17 cells. Nat Immunol 2016(17):1084–1092

Chen YJ, Ho HJ, Tseng CH, Lai ZL, Shieh JJ, Wu CY (2018) Intestinal microbiota profiling and predicted metabolic dysregulation in psoriasis patients. Exp Dermatol 27:1336–1343

Christou EAA, Banos A, Kosmara D, Bertsias GK, Boumpas DT (2019) Sexual dimorphism in SLE: above and beyond sex hormones. Lupus 28:3–10

Chu H, Khosravi A, Kusumawardhani IP, Kwon AH, Vasconcelos AC, Cunha LD et al (2016) Gene–microbiota interactions contribute to the pathogenesis of inflammatory bowel disease. Science 352:1116–1120

Chua HH, Chou HC, Tung YL, Chiang BL, Liao CC, Liu HH et al (2018) Intestinal dysbiosis featuring abundance of Ruminococcus gnavus associates with allergic diseases in infants. Gastroenterology 154:154–167

Clayton K, Vallejo A, Sirvent S, Davies J, Porter G, Reading IC et al (2020) Machine learning applied to atopic dermatitis transcriptome reveals distinct therapy-dependent modification of the keratinocyte immunophenotype. Br J Dermatol 184(5):913–922. https://doi.org/10.1111/bjd.19431. Epub 2020 Sep 20. PMID: 32730675

Codoñer FM, Ramírez-Bosca A, Climent E, Carrión-Gutierrez M, Guerrero M, Pérez-Orquín JM, Horga de la Parte J, Genovés S, Ramón D, Navarro-López V, Chenoll E (2018) Gut microbial composition in patients with psoriasis. Sci Rep 8(1):3812. https://doi.org/10.1038/s41598-018-22125-y. PMID: 29491401; PMCID: PMC5830498

Colucci R, Moretti S (2021) Implication of human bacterial gut microbiota on immune-mediated and autoimmune dermatological diseases and their comorbidities: a narrative review. Dermatol Ther 11:363–384. https://doi.org/10.1007/s13555-021-00485-0

Cording S, Fleissner D, Heimesaat MM, Bereswill S, Loddenkemper C, Uematsu S et al (2013) Commensal microbiota drive proliferation of conventional and Foxp3(+) regulatory CD4(+) T cells in mesenteric lymph nodes and Peyer's patches. Eur J Microbiol Immunol 3:1–10

Couturier-Maillard A, Secher T, Rehman A, Normand S, De Arcangelis A, Haesler R et al (2013) NOD2-mediated dysbiosis predisposes mice to transmissible colitis and colorectal cancer. J Clin Invest 123:700–711

Cullender TC, Chassaing B, Janzon A, Kumar K, Muller CE, Werner JJ et al (2013) Innate and adaptive immunity interact to quench microbiome flagellar motility in the gut. Cell Host Microbe 14:571–581

De Santis S, Cavalcanti E, Mastronardi M, Jirillo E, Chieppa M (2015) Nutritional keys for intestinal barrier modulation. Front Immunol 6:612

Dominguez-Bello MG, Costello EK, Contreras M, Magris M, Hidalgo G, Fierer N et al (2010) Delivery mode shapes the acquisition and structure of the initial microbiota across multiple body habitats in newborns. Proc Natl Acad Sci U S A 107:11971–11975

Eri T, Kawahata K, Kanzaki T, Imamura M, Michishita K, Akahira L et al (2017) Intestinal microbiota link lymphopenia to murine autoimmunity via PD-1(+)CXCR5(−/dim) B-helper T cell induction. Sci Rep 7:46037

Fagarasan S, Kawamoto S, Kanagawa O, Suzuki K (2010) Adaptive immune regulation in the gut: T cell-dependent and T cell-independent IgA synthesis. Annu Rev Immunol 28:243–273

Feng T, Wang L, Schoeb TR, Elson CO, Cong Y (2010) Microbiota innate stimulation is a prerequisite for T cell spontaneous proliferation and induction of experimental colitis. J Exp Med 207:1321–1332
Floreani A, Leung PS, Gershwin ME (2016) Environmental basis of autoimmunity. Clin Rev Allergy Immunol 50:287–300
Fransen F, Zagato E, Mazzini E, Fosso B, Manzari C, El Aidy S et al (2015) BALB/c and C57BL/6 mice differ in polyreactive IgA abundance, which impacts the generation of antigen-specific IgA and microbiota diversity. Immunity 43:527–540
Furusawa Y, Obata Y, Fukuda S, Endo TA, Nakato G, Takahashi D et al (2013) Commensal microbe-derived butyrate induces the differentiation of colonic regulatory T cells. Nature 504: 446–450
Gaboriau-Routhiau V, Rakotobe S, Lecuyer E, Mulder I, Lan A, Bridonneau C et al (2009) The key role of segmented filamentous bacteria in the coordinated maturation of gut helper T cell responses. Immunity 2009(31):677–689
Gittler JK, Shemer A, Suárez-Fariñas M, Fuentes Duculan J, Gulewicz KJ, Wang CQ et al (2012) Progressive activation of T(H)2/T(H)22 cytokines and selective epidermal proteins characterizes acute and chronic atopic dermatitis. J Allergy Clin Immunol 130:1344–1354
Goto Y, Panea C, Nakato G, Cebula A, Lee C, Diez MG et al (2014) Segmented filamentous bacteria antigens presented by intestinal dendritic cells drive mucosal Th17 cell differentiation. Immunity 2014(40):594–607
Grän F, Kerstan A, Serfling E, Goebeler M, Muhammad K (2020) Current developments in the immunology of psoriasis. Yale J Biol Med 93:97–110
Grashoff C, Hoffman BD, Brenner MD, Zhou R, Parsons M, Yang MT, McLean MA, Sligar SG, Chen CS, Ha T, Schwartz MA (2010) Measuring mechanical tension across vinculin reveals regulation of focal adhesion dynamics. Nature 466(7303):263–266. https://doi.org/10.1038/nature09198. PMID: 20613844; PMCID: PMC2901888
Graßhoff H, Comdühr S, Monne LR, Müller A, Lamprecht P, Riemekasten G, Humrich JY (2021) Low-dose IL-2. Therapy in autoimmune and rheumatic diseases. Front Immunol 12:648408. https://doi.org/10.3389/fimmu.2021.648408
Greiling TM, Dehner C, Chen X, Hughes K, Iniguez AJ, Boccitto M et al (2018) Commensal orthologs of the human autoantigen Ro60 as triggers of autoimmunity in lupus. Sci Transl Med 10:eaan2306
He Z, Shao T, Li H, Xie Z, Wen C (2016) Alterations of the gut microbiome in Chinese patients with systemic lupus erythematosus. Gut Pathogens 8:64. https://doi.org/10.1186/s13099-016-0146-9
Hevia A, Milani C, Lopez P, Cuervo A, Arboleya S, Duranti S et al (2014) Intestinal dysbiosis associated with systemic lupus erythematosus. MBio 5:e01548–ee1614
Hidalgo-Cantabrana C, Gomez J, Delgado S, Requena-Lopez S, Queiro-Silva R, Margolles A et al (2019) Gut microbiota dysbiosis in a cohort of patients with psoriasis. Br J Dermatol 181:1287–1295
Ho NT, Li F, Lee-Sarwar KA, Tun HM, Brown BP, Pannaraj PS et al (2018) Meta-analysis of effects of exclusive breastfeeding on infant gut microbiota across populations. Nat Commun 9: 4169
Hong P-Y, Lee BW, Aw M, Shek LPC, Yap GC, Chua KY et al (2010) Comparative analysis of fecal microbiota in infants with and without eczema. PLoS One 5:e9964
Horai R, Zarate-Blades CR, Dillenburg-Pilla P, Chen J, Kielczewski JL, Silver PB et al (2015) Microbiota-dependent activation of an autoreactive T cell receptor provokes autoimmunity in an immunologically privileged site. Immunity 43:343–353
Horta-Baas G, Romero-Figueroa MDS, Montiel-Jarquín AJ, Pizano-Zárate ML, García-Mena J, Ramírez-Durán N (2017) Intestinal dysbiosis and rheumatoid arthritis: a link between gut microbiota and the pathogenesis of rheumatoid arthritis. J Immunol Res 2017:4835189. Epub 2017 Aug 30. PMID: 28948174; PMCID: PMC5602494. https://doi.org/10.1155/2017/4835189

Horz HP (2015) Archaeal lineages within the human microbiome: absent, rare or elusive? Life (Basel) 5:1333–1345

Hu Y, Costenbader KH, Gao X, Hu FB, Karlson EW, Lu B (2015) Mediterranean diet and incidence of rheumatoid arthritis in women. Arthritis Care Res 67(5):597–606. https://doi.org/10.1002/acr.2248

Huang L, Gao R, Yu N, Zhu Y, Ding Y, Qin H (2019) Dysbiosis of gut microbiota was closely associated with psoriasis. Sci China Life Sci 62:807–815

Humrich JY, Riemekasten G (2016a) Restoring regulation—IL-2 therapy in systemic lupus erythematosus. Expert Rev Clin Immunol 12:1153–1160. https://doi.org/10.1080/1744666X.2016.1199957

Humrich JY, Riemekasten G (2016b) Clinical trials: the rise of IL-2 therapy - a novel biologic treatment for SLE. Nat Rev Rheumatol 12:695–696. https://doi.org/10.1038/nrrheum.2016.173

Isailovic N, Daigo K, Mantovani A, Selmi C (2015) Interleukin-17 and innate immunity in infections and chronic inflammation. J Autoimmun 60:1–11

Johnsona BM, Gaudreaua MC, Gudia R, Browna R, Gilkesonb G, Vasua C (2020) Gut microbiota differently contributes to intestinal immune phenotype and systemic autoimmune progression in female and male lupus-prone mice. J Autoimmun 108:102420

Jones JL, Thompson SA, Loh P, Davies JL, Tuohy OC, Curry AJ et al (2013) Human autoimmunity after lymphocyte depletion is caused by homeostatic T-cell proliferation. Proc Natl Acad Sci U S A 110:20200–20205

Kapur S, Watson W, Carr S (2018) Atopic dermatitis. Allergy asthma. Clin Immunol 14:52

Kawamoto S, Maruya M, Kato LM, Suda W, Atarashi K, Doi Y et al (2014) Foxp3(+) T cells regulate immunoglobulin a selection and facilitate diversification of bacterial species responsible for immune homeostasis. Immunity 41:152–165

Kim JK, Kwok SK, Choe JY, Park SH (2019) Recent advances in our understanding of the link between the intestinal microbiota and systemic lupus erythematosus. Int J Mol Sci 20:4871

Kjeldsen-Kragh J, Borchgrevink CF, Laerum E et al (1991) Controlled trial of fasting and one-year vegetarian diet in rheumatoid arthritis. Lancet 338(8772):899–902

Klatzmann D, Abbas AK (2015) The promise of low-dose interleukin-2 therapy for autoimmune and inflammatory diseases. Nat Rev Immunol 15:283–294. https://doi.org/10.1038/nri3823

Kocic H, Damiani G, Stamenkovic B, Tirant M, Jovic A, Tiodorovic D et al (2019) Dietary compounds as potential modulators of microRNA expression in psoriasis. Ther Adv Chronic Dis 10:2040622319864805

Kosiewicz MM, Dryden GW, Chhabra A, Alard P (2014) Relationship between gut microbiota and development of T cell associated disease. FEBS Lett 588:4195–4206

Krasselt M, Baerwald C (2019) Sex, symptom severity, and quality of life in rheumatology. Clin Rev Allergy Immunol 56:346–361

Kuhn A, Bonsmann G, Anders HJ, Herzer P, Tenbrok K, Schneider M (2015) The diagnosis and treatment of systemic lupus erythematosus. Dtsch Arztebl Int 112:423–432

Laurence A, Tato CM, Davidson TS, Kanno Y, Chen Z, Yao Z et al (2007) Interleukin-2 signaling via STAT5 constrains T helper 17 cell generation. Immunity 26:371–381. https://doi.org/10.1016/j.immuni.2007.02.009

Laursen MF, Zachariassen G, Bahl MI, Bergström A, Høst A, Michaelsen KF et al (2015) Having older siblings is associated with gut microbiota development during early childhood. BMC Microbiol 15:154

Lecuyer E, Rakotobe S, Lengline-Garnier H, Lebreton C, Picard M, Juste C et al (2014) Segmented filamentous bacterium uses secondary and tertiary lymphoid tissues to induce gut IgA and specific T helper 17 cell responses. Immunity 40:608–620

Lee YK, Menezes JS, Umesaki Y, Mazmanian SK (2011) Proinflammatory T-cell responses to gut microbiota promote experimental autoimmune encephalomyelitis. Proc Natl Acad Sci U S A 108(Suppl 1):4615–4622

Lee MJ, Kang MJ, Lee SY, Lee E, Kim K, Won S et al (2018a) Perturbations of gut microbiome genes in infants with atopic dermatitis according to feeding type. J Allergy Clin Immunol 141: 1310–1319
Lee S-Y, Lee E, Park YM, Hong A-J (2018b) Microbiome in the gut-skin axis in atopic dermatitis. Allergy Asthma Immunol Res 2018(10):354–362
Lepe K, Zito PM (2020) Alopecia Areata. In: StatPearls [Internet]. StatPearls, Treasure Island, FL
Li M, Wang F (2021) Review article role of intestinal microbiota on gut homeostasis and rheumatoid arthritis. J Immunol Res 2021:8167283., 9 pages. https://doi.org/10.1155/2021/8167283
Li YY, Pearson JA, Chao C, Peng J, Zhang X, Zhou Z et al (2017) Nucleotidebinding oligomerization domain-containing protein 2 (Nod2) modulates T1DM susceptibility by gut microbiota. J Autoimmun 82:85–95
Li Y, Wang HF, Li X, Li HX, Zhang Q, Zhou HW et al (2019) Disordered intestinal microbes are associated with the activity of systemic lupus erythematosus. Clin Sci (Lond) 133:821–838
Lin L, Zhang J (2017) Role of intestinal microbiota and metabolites on gut homeostasis and human diseases. BMC Immunol 18(1):2. PMID: 28061847; PMCID: PMC5219689. https://doi.org/10.1186/s12865-016-0187-3
Lloyd-Price J, Abu-Ali G, Curtis HC (2016) The healthy human microbiome. Genome Med 8:51
López P, de Paz B, Rodríguez-Carrio J, Hevia A, Sánchez B, Margolles A et al (2016) Th17 responses and natural IgM antibodies are related to gut microbiota composition in systemic lupus erythematosus patients. Sci Rep 6:24072
Luo XM, Edwards MR, Mu Q, Yu Y, Vieson MD, Reilly CM et al (2018) Gut microbiota in human systemic lupus erythematosus and a mouse model of lupus. Appl Environ Microbiol 84: e02288–ee2317
Lynch SV, Pedersen O (2016) The human intestinal microbiome in health and disease. N Engl J Med 375(24):2369–2379
Madsen AL, Schack-Nielsen L, Larnkjaer A, Mølgaard C, Michaelsen KF (2010) Determinants of blood glucose and insulin in healthy 9-month-old term Danish infants; the SKOT cohort. Diabet Med 27:1350–1357
Mahdavinia M, Rasmussen HE, Engen P, Van den Berg JP, Davis E, Engen K et al (2017) Atopic dermatitis and food sensitization in south African toddlers: role of fiber and gut microbiota. Ann Allergy Asthma Immunol 118:742–743.e3
Malek TR (2008) The biology of interleukin-2. Annu Rev Immunol 26:453–479. https://doi.org/10.1146/annurev.immunol.26.021607.090357
Malek TR, Castro I (2010) Interleukin-2 receptor signaling: at the interface between tolerance and immunity. Immunity 33:153–165. https://doi.org/10.1016/j.immuni.2010.08.004
Markle JCM, Frank DN, Mortin-Toth S, Robertson CE, Feazel LM, Rolle-Kampczyk U et al (2013) Sex differences in the gut microbiome drive hormone-dependent regulation of autoimmunity. Science 339:1084–1088
Megna M, Patruno C, Balato A, Rongioletti F, Stingeni L, Balato N (2017) Italian adult atopic dermatitis study group. An Italian multicentre study on adult atopic dermatitis: persistent versus adult-onset disease. Arch Dermatol Res 309:443–452
Miyara M, Gorochov G, Ehrenstein M, Musset L, Sakaguchi S, Amoura Z (2011) Human FoxP3+ regulatory T cells in systemic autoimmune diseases. Autoimmun Rev 10:744–755. https://doi.org/10.1016/j.autrev.2011.05.004
Moon C, Baldridge MT, Wallace MA, Burnham CA, Virgin HW, Stappenbeck TS (2015) Vertically transmitted faecal IgA levels determine extra-chromosomal phenotypic variation. Nature 521:90–93
Moreno-Arrones OM, Serrano-Villar S, Perez-Brocal V, Saceda-Corralo D, Morales-Raya C, Rodrigues- Barata R et al (2020) Analysis of the gut microbiota in alopecia areata: identification of bacterial biomarkers. J Eur Acad Dermatol Venereol 34:400–405
Morrison DJ, Preston T (2016) Formation of short chain fatty acids by the gut microbiota and their impact on human metabolism. Gut Microbes 7:189–200

Mu Q, Kirby J, Reilly CM, Luo XM (2017) Leaky gut as a danger signal for autoimmune diseases. Front Immunol 8:598. https://doi.org/10.3389/fimmu.2017.00598

Nair PA, Badri T (2020) Psoriasis. In: StatPearls [internet]. StatPearls, Treasure Island, FL

Nakahara T, Kido-Nakahara M, Tsuji G, Furue M (2020) Basics and recent advances in the pathophysiology of atopic dermatitis. J Dermatol 48(2):130–139. https://doi.org/10.1111/1346-8138.15664

Nielen MM, van Schaardenburg D, Reesink HW et al (2004) Specific autoantibodies precede the symptoms of rheumatoid arthritis: a study of serial measurements in blood donors. Arthritis Rheum 50(2):380–386

Nielsen PR, Kragstrup TW, Deleuran BW, Benros ME (2016) Infections as risk factor for autoimmune diseases—a nationwide study. J Autoimmun 74:176–181

O'Dell JR, Blakely KW, Mallek JA, Eckhoff PJ, Leff RD, Wees SJ, Sems KM, Fernandez AM, Palmer WR, Klassen LW, Paulsen GA, Haire CE, Moore GF (2001) Treatment of early seropositive rheumatoid arthritis: a two-year, double-blind comparison of minocycline and hydroxychloroquine. Arthritis Rheum 44(10):2235–2241. PMID: 11665963. https://doi.org/10.1002/1529-0131(200110)44:10<2235::aid-art385>3.0.co;2-a

Ohnmacht C, Park JH, Cording S, Wing JB, Atarashi K, Obata Y et al (2015) Mucosal immunology: the microbiota regulates type 2 immunity through RORgammat(+) T cells. Science 349:989–993

Ouwerkerk JP, de Vos WM, Belzer B (2013) Glycobiome: bacteria and mucus at the epithelial interface. Best Pract Res Clin Gastroenterol 27:25–38

Palm NW, de Zoete MR, Cullen TW, Barry NA, Stefanowski J, Hao L et al (2014) Immunoglobulin a coating identifies colitogenic bacteria in inflammatory bowel disease. Cell 158:1000–1010

Panea C, Farkas AM, Goto Y, Abdollahi-Roodsaz S, Lee C, Koscso B et al (2015) Intestinal monocyte-derived macrophages control commensal specific Th17 responses. Cell Rep 12:1314–1324

Penders J, Stobberingh EE, van den Brandt PA, Thijs C (2007a) The role of the intestinal microbiota in the development of atopic disorders. Allergy 62:1223–1236

Penders J, Thijs C, van den Brandt PA, Kummeling I, Snijders B, Stelma F, Adams H, van Ree R, Stobberingh EE (2007b) Gut microbiota composition and development of atopic manifestations in infancy: the KOALA Birth Cohort Study. Gut 56(5):661–667. https://doi.org/10.1136/gut.2006.100164. Epub 2006 Oct 17. PMID: 17047098; PMCID: PMC1942165

Petersen EBM, Skov L, Thyssen JP, Jensen P (2019) Role of the gut microbiota in atopic dermatitis: a systematic review. Acta Derm Venereol 99:5–11

Polkowska-Pruszyńska B, Gerkowicz A, Krasowska D (2020) The gut microbiome alterations in allergic and inflammatory skin diseases—an update. J Eur Acad Dermatol Venereol 34:455–464

Qin J, Li R, Raes J, Arumugam M, Solvsten K et al (2010) A human gut microbial gene catalogue established by metagenomic sequencing. Nature 464:59–65

Raczkiewicz A, Kisiel B, Kulig M, Tłustochowicz W (2015) Vitamin D status and its association with quality of life, physical activity, and disease activity in rheumatoid arthritis patients. J Clin Rheumatol 21(3):126–130. https://doi.org/10.1097/RHU.0000000000000233

Ramanan D, Tang MS, Bowcutt R, Loke P, Cadwell K (2014) Bacterial sensor Nod2 prevents inflammation of the small intestine by restricting the expansion of the commensal Bacteroides vulgatus. Immunity 41:311–324

Ramírez-Boscá A, Navarro-López V, Martínez-Andrés A, Such J, Francés R, Horga de la Parte J et al (2015) Identification of bacterial DNA in the peripheral blood of patients with active psoriasis. JAMA Dermatol 151:670–671

Rea D, Coppola G, Palma G, Barbieri A, Luciano A, Del Prete P et al (2018) Microbiota effects on cancer: from risks to therapies. Oncotarget 9:17915–17927

Rebello D, Wang E, Yen E, Lio PA, Kelly CR (2017) Hair growth in two alopecia patients after fecal microbiota transplant. ACG Case Rep J 2017(4):e107

Reddel S, Del Chierico F, Quagliariello A et al (2019) Gut microbiota profile in children affected by atopic dermatitis and evaluation of intestinal persistence of a probiotic mixture. Sci Rep 9:4996. https://doi.org/10.1038/s41598-019-41149-6

Ritchlin CT, Colbert RA, Gladman DD (2017) Psoriatic arthritis. N Engl J Med 376:957–970
Rodríguez-Carrio J, López P, Sánchez B, González S, Gueimonde M, Margolles A et al (2017) Intestinal dysbiosis is associated with altered short-chain fatty acids and serum-free fatty acids in systemic lupus erythematosus. Front Immunol 8:23
Round JL, O'Connell RM, Mazmanian SK (2010) Coordination of tolerogenic immune responses by the commensal microbiota. J Autoimmun 34:J220–J225
Sacotte R, Silverberg JI (2018) Epidemiology of adult atopic dermatitis. Clin Dermatol 36:595–605
Sakaguchi S (2005) Naturally arising Foxp3-expressing CD25+CD4+ regulatory T cells in immunological tolerance to self and non-self. Nat Immunol 6:345–352. https://doi.org/10.1038/ni1178
Sakaguchi S, Miyara M, Costantino CM, Hafler DA (2010) FOXP3+ regulatory T cells in the human immune system. Nat Rev Immunol 10:490–500. https://doi.org/10.1038/nri2785
Salem I, Ramser A, Isham N, Ghannoum MA (2018) The gut microbiome as a major regulator of the gut-skin axis. Front Microbiol 9:1459
Sano T, Huang W, Hall JA, Yang Y, Chen A, Gavzy SJ et al (2015) An IL-23R/IL-22 circuit regulates epithelial serum amyloid a to promote local effector Th17 responses. Cell 163:381–393
Scarpellini E, Ianiro G, Attili F, Bassanelli C, De Santis A, Gasbarrini A (2015) The human gut microbiota and virome: potential therapeutic implications. Dig Liver Dis 47:1007–1012
Scher JU, Ubeda C, Artacho A, Attur M, Isaac S, Reddy SM et al (2015) Decreased bacterial diversity characterizes an altered gut microbiota in psoriatic arthritis and resembles dysbiosis of inflammatory bowel disease. Arthritis Rheumatol 67:128–139
Schiering C, Krausgruber T, Chomka A, Frohlich A, Adelmann K, Wohlfert EA et al (2014) The alarmin IL-33 promotes regulatory T-cell function in the intestine. Nature 513:564–568
Selber-Hnatiw S, Rukundo B, Ahmadi M, Akoubi H, Al-Bizri H, Aliu AF (2017) Human gut microbiota toward an ecology of disease. Front Microbiol 8:1264
Shapiro J, Cohen NA, Shalev V, Uzan A, Koren O, Maharshak N (2019) Psoriatic patients have a distinct structural and functional fecal microbiota compared with controls. J Dermatol 46:595–603
Sharabi A, Tsokos MG, Ding Y, Malek TR, Klatzmann D, Tsokos GC (2018) Regulatory T cells in the treatment of disease. Nat Rev Drug Discov 17:823–844. https://doi.org/10.1038/nrd.2018.148
Sikora M, Stec A, Chrabaszcz M, Waskiel-Burnat A, Zaremba M, Olszewska M et al (2019) Intestinal fatty acid binding protein, a biomarker of intestinal barrier, is associated with severity of psoriasis. J Clin Med 2(8):1021
Silverman GJ, Azzouz DF, Alekseyenko AV (2019) Systemic lupus erythematosus and dysbiosis in the microbiome: cause or effect or both? Curr Opin Immunol 61:80–85
Smith PM, Howitt MR, Panikov N, Michaud M, Gallini CA, Bohlooly YM et al (2013) The microbial metabolites, short-chain fatty acids, regulate colonic Treg cell homeostasis. Science 341:569–573
Tan TG, Sefik E, Geva-Zatorsky N, Kua L, Naskar D, Teng F et al (2016) Identifying species of symbiont bacteria from the human gut that, alone, can induce intestinal Th17 cells in mice. Proc Natl Acad Sci U S A 113:E8141–E8150
Tan L, Zhao S, Zhu W, Wu L, Li J, Shen M et al (2018) The Akkermansia muciniphila is a gut microbiota signature in psoriasis. Exp Dermatol 27:144–149
Tanoue T, Atarashi K, Honda K (2016) Development and maintenance of intestinal regulatory T cells. Nat Rev Immunol 16:295–309
Telesford KM, Yan W, Ochoa-Reparaz J, Pant A, Kircher C, Christy MA et al (2015) A commensal symbiotic factor derived from Bacteroides fragilis promotes human CD39(+)Foxp3(+) T cells and Treg function. Gut Microbes 6:234–242
Thursby E, Juge N (2017) Introduction to the human gut microbiota. Biochem J 474:1823–1836
Tsokos GC (2020) Autoimmunity and organ damage in systemic lupus erythematosus. Nat Immunol 21:605–614. https://doi.org/10.1038/s41590-020-0677-6

Van der Meulen TA, Harmsen HJM, Vila AV, Kurilshikov A, Liefers SC, Zhernakova A et al (2019) Shared gut, but distinct oral microbiota composition in primary Sjogren's syndrome and systemic lupus erythematosus. J Autoimmun 97:77–87

Van Praet JT, Donovan E, Vanassche I, Drennan MB, Windels F, Dendooven A et al (2015) Commensal microbiota influence systemic autoimmune responses. EMBO J 34:466–474

Vatanen T, Kostic AD, d'Hennezel E, Siljander H, Franzosa EA, Yassour M et al (2016) Variation in microbiome LPS immunogenicity contributes to autoimmunity in humans. Cell 165:842–853

Vemuri R, Sylvia KE, Klein SL, Forster SC, Plebanski M, Eri R, Flanagan KL (2019) The microgenderome revealed: sex differences in bidirectional interactions between the microbiota, hormones, immunity and disease susceptibility. Semin Immunopathol 41(2):265–275. https://doi.org/10.1007/s00281-018-0716-7. Epub 2018 Oct 8. PMID: 30298433; PMCID: PMC6500089

Vijay-Kumar M, Aitken JD, Carvalho FA, Cullender TC, Mwangi S, Srinivasan S et al (2010) Metabolic syndrome and altered gut microbiota in mice lacking toll-like receptor 5. Science 328: 228–231

Viladomiu M, Kivolowitz C, Abdulhamid A, Dogan B, Victorio D, Castellanos JG et al (2017) IgA-coated E. coli enriched in Crohn's disease spondyloarthritis promote TH17-dependent inflammation. Sci Transl Med 9:376

Visser MJ, Kell DB, Pretorius E et al (2019) Bacterial dysbiosis and translocation in psoriasis vulgaris. Front Cell Infect Microbiol 9:7

von Spee-Mayer C, Siegert E, Abdirama D, Rose A, Klaus A, Alexander T et al (2016) Low-dose interleukin-2 selectively corrects regulatory T cell defects in patients with systemic lupus erythematosus. Ann Rheum Dis 75:1407–1415. https://doi.org/10.1136/annrheumdis-2015-207776

Wang Y, Telesford KM, Ochoa-Reparaz J, Haque-Begum S, Christy M, Kasper EJ et al (2014) An intestinal commensal symbiosis factor controls neuroinflammation via TLR2-mediated CD39 signalling. Nat Commun 5:4432

Wei F, Xu H, Yan C, Rong C, Liu B, Zhou H (2019) Changes of intestinal flora in patients with systemic lupus erythematosus in Northeast China. PLoS One 14:e0213063

Wen L, Ley RE, Volchkov PY, Stranges PB, Avanesyan L, Stonebraker AC et al (2008) Innate immunity and intestinal microbiota in the development of type 1 diabetes. Nature 455:1109–1113

West CE, Ryden P, Lundin D, Engstrand L, Tulic MK, Prescott SL (2015) Gut microbiome and innate immune response patterns in IgE-associated eczema. Clin Exp Allergy 45:1419–1429

Wilck N, Matus MG, Kearney SM, Olesen SW, Forslund K, Bartolomaeus H et al (2017) Salt-responsive gut commensal modulates TH17 axis and disease. Nature 551:585–589

Wu HJ, Ivanov II, Darce J, Hattori K, Shima T, Umesaki Y et al (2010) Gutresiding segmented filamentous bacteria drive autoimmune arthritis via T helper 17 cells. Immunity 32:815–827

Xie WR, Yang XY, Xia HH, Wu LH, He XX (2019) Hair regrowth following fecal microbiota transplantation in an elderly patient with alopecia areata: a case report and review of the literature. World J Clin Cases 7:3074–3081

Yang Y, Torchinsky MB, Gobert M, Xiong H, Xu M, Linehan JL et al (2014) Focused specificity of intestinal TH17 cells towards commensal bacterial antigens. Nature 510:152–156

Yang BH, Hagemann S, Mamareli P et al (2016) (2016). Foxp3+ T cells expressing RORγt represent a stable regulatory T-cell effector lineage with enhanced suppressive capacity during intestinal inflammation. Mucosal Immunol 9:444–457. https://doi.org/10.1038/mi.2015.74

Yang L, Fu J, Zhou Y (2020) Research progress in atopic march. Front Immunol 11:1907

Zechner EL (2017) Inflammatory disease caused by intestinal pathobionts. Curr Opin Microbiol 35: 64–69

Zheng H, Liang H, Wang Y, Miao M, Shi T, Yang F (2016) Altered gut microbiota composition associated with eczema in infants. PLoS One 11:e0166026

References

Our Perception of Autoimmunity and Microbes

7

7.1 Introduction

This chapter focuses on our perceptions and hypotheses about autoimmunity based on our previous studies.

7.2 The Expression of Estrogen Receptor and Bcl2 in *Candida albicans* May Represent Removal of Functional Barriers Among Eukaryotic and Prokaryotic Cells

Even though there are numerous anatomical differences between prokaryotic and eukaryotic cells, we believe that functional similarities can lead to differences. The link between estrogen exposure and *Candida albicans* pathogenicity has been studied for a long time. The purpose of this study was to look at the expression of estrogen receptor and BCL2 in *Candida albicans. Candida albicans* (ATCC) was cultured for 24 h at 37 °C in Sabouraud Dextrose Broth (SDB). For 5 min, a tube containing *C. albicans* and broth medium (SDB) was centrifuged to concentrate the amount of *C. albicans* (3000 RPM). To generate *C. albicans* smears, a sample of the sediment was transferred to slides. The estrogen receptor (ER) and BCL2 proteins were immunohistochemically localized utilizing indirect immunoperoxide methods.

In *C. albicans*, both the ER and BCL2 proteins were found. *C. albicans* localization was found primarily in the nucleus and to a lesser extent in the cytoplasm, and in the hyphae form of *C. albicans*. Taken together, our preliminary findings suggest that the effects of estrogen and its receptor on *C. albicans* are more complex than previously thought, and that *C. albicans* growth may be governed by mechanisms similar to those found in host cells, which will help us better understand how prokaryotic and eukaryotic cells interact (Alkhatib 2017).

A. J. Alkhatib, *The Role of Microbes in Autoimmune Diseases*,
https://doi.org/10.1007/978-981-19-1162-0_7

7.3 Expression and Releasing of Cell Cycle Proteins by *Candida albicans* into Surrounding Tissue: New Perspectives of the Relationship Between Microbes and Host

The goal of this research is to give new philosophical viewpoints on how *Candida albicans* (*C. albicans*) interacts with host cells. We rely on a novel technique in which *C. albicans* produces and releases key proteins like the estrogen receptor. Immunohistochemical techniques were used to investigate this method. *C. albicans* from nutrition broth medium was administered intraperitoneally into rats. Rats were murdered using anesthetic procedures 2 weeks later. The findings demonstrated that ER is expressed and released in *C. albicans* in both the cellular and hyphal forms. The presence of ER as an exudate from infection sites towards host cells was discovered. Taken together, this is the first study to disclose the interaction between *C. albicans* and host cells through the expression and release of ER, which may impede host cell stability and induction of disease development and progression, to the best of the researcher's knowledge. More sophisticated approaches may be required to confirm the findings of our investigation (Alkhatib 2019).

7.4 Non-classical Roles of Microbes

Microorganisms' traditional roles include things like being involved in disease etiology; a vast number of diseases are known to be caused by microbes (Costerton et al. 1999; Autenrieth 2017; Wang et al. 2018). Adherence and competitiveness with dangerous microorganisms are examples of other positive activities (Canny and McCormick 2008; Hibbing et al. 2010). Microbes have also been used in the industrial world in a variety of fields, including genetic engineering (Godwill 2014). In fact, the roles of microorganisms have other dimensions that may broaden our understanding of our interior world (Alkhatib 2018). I'd want to share some of our expertise in the field of microbiology with reference to bacteria non-classical roles. Using ELISA and PCR, we discovered that roughly half of the patients with cardiac issues who were indicated for catheterization were positive for *Chlamydia pneumonia* (Al-khatib and Al-Alawneh 2013). Our results, together with others of a similar nature, have the potential to transform the treatment of cardiac disease. Microbes also play a non-traditional function in crime. We analyzed 1000 inmates in several Jordanian prisons and discovered a link between crime and latent toxoplasmosis. In this setting, bacteria have an impact on our mental health and interfere with our vision of life (Shotar et al. 2015a, b, 2016). The perception of good and terrible is not solely based on social factors. Microbes have an indirect impact on our values and ethics. As their products circulate in the blood and affect our neurological system, I dubbed these microorganisms "micro-evils" in this context (Wexler 2007; Gandon and Vale 2014). I recently discovered that *Candida albicans* expresses a number of important proteins, including estrogen receptor and BCL2, implying that the pathogenicity of *Candida albicans* is dependent on these proteins on one hand, and that functional barriers between microbes and host cells may be broken down, resulting in new concepts of microbe-host cell interaction (Alkhatib 2017).

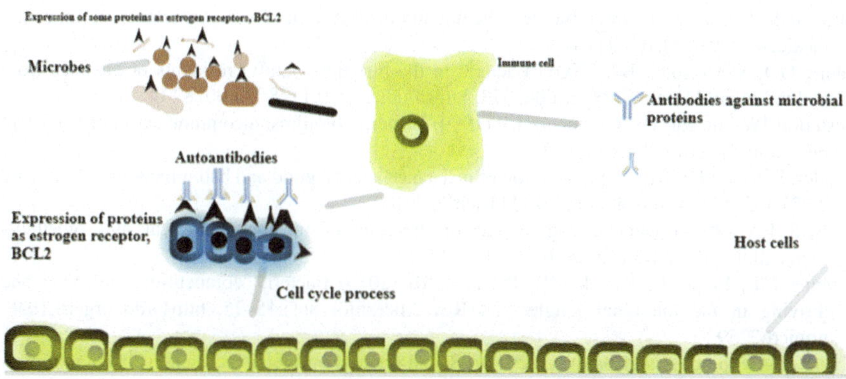

Fig. 7.1 Schematic diagram of proposed model of autoimmunity initiation

7.5 Proposed Model Explaining the Initiation of Autoimmunity by Microbes

It is clear to us, those similar components are shared among microbes and host cells. The problem involved traditionally is that such proteins are conserved, which implies that immune cells are not recognizing them as foreign bodies. The results of our studies pointed to the exposure of self-antigens as foreign antigens by mimicry mechanisms through presenting them by microbes. As far as the immune cells are exposed to these antigens, an immunologic interaction is likely to occur. It is worth mentioning that the high homogeneity involved between microbial antigens and host antigens generates autoantibodies that will hit self-antigens, or they precipitate in various body tissues and organs including pancreas leading to diabetes, endocarditis, pyelonephritis, or neurological disorders including multiple sclerosis.

Based on the consideration that billions of microbes exist in our bodies, particularly the intestine, similarity of antigens burdens the immune system leading to what I call "genetic crisis as a new possible syndrome" (Alkhatib 2021) (Fig. 7.1).

References

Alkhatib AJ (2017) The expression of estrogen receptor and Bcl2 in *Candida albicans* may represent removal of functional barriers among eukaryotic and prokaryotic cells. EC Microbiol SI.01:20–23

Alkhatib AJ (2018) Non-classical roles of microbes. PSM Microbiol 3(2):62–63

Alkhatib AJ (2019) Expression and releasing of cell cycle proteins by *Candida albicans* into surrounding tissue: new perspectives of the relationship between microbes and host. EC Microbiol 15(3):168–171

Alkhatib AJ (2021) Genetic crisis as a new possible syndrome. Acta Sci Clin Case Rep 2(3):30–31

Al-khatib AJ, Al-Alawneh M (2013) Exploring the relationship between the infection of C. pneumoniae and coronary artery disease. Eur Sci J 9(6):195–213

Autenrieth IB (2017) The microbiome in health and disease: a new role of microbes in molecular medicine. J Mol Med 95(1):1–3
Canny GO, McCormick BA (2008) Bacteria in the intestine, helpful residents or enemies from within? Infect Immun 76(8):3360–3373. https://doi.org/10.1128/IAI.00187-08
Costerton JW, Stewart PS, Greenberg EP (1999) Bacterial biofilms: a common cause of persistent infections. Science 284:1318–1322
Gandon S, Vale PF (2014) The evolution of resistance against good and bad infections. J Evol Biol 27(2):303–312. https://doi.org/10.1111/jeb.12291
Godwill EA (2014) Genetic engineering on microorganism: the ecological and bioethical implications. Eur J Biotechnol Biosci 1(3):27–33
Hibbing ME, Fuqua C, Parsek MR, Peterson SB (2010) Bacterial competition: surviving and thriving in the microbial jungle. Nat Rev Microbiol 8(1):15–25. https://doi.org/10.1038/nrmicro2259
Shotar A, Alzyoud SA, AlKhatib AJ (2015a) Social impacts of infectious diseases: latent toxoplasmosis and crime. Soc Sci 10:1677–1681
Shotar A, Alzyoud SA, AlKhatib AJ (2015b) The impacts of latent toxoplasmosis on physical violent actions among a sample of Jordanian inmates. Res J Biol Sci 10:72–77
Shotar A, Alzyoud SA, AlKhatib AJ (2016) Latent toxoplasmosis and the involvement in road traffic accidents among a sample of Jordanian drivers. Res J Med Sci 10:194–198
Wang HL, Cheng X, Ding SW, Wang DW, Chen C, Xu CL, Xie H (2018) Molecular identification and functional characterization of the cathepsin B gene (Ab-cb-1) in the plant parasitic nematode Aphelenchoides besseyi. PLoS One 13(6):e0199935. https://doi.org/10.1371/journal.pone.0199935
Wexler HM (2007) Bacteroides: the good, the bad, and the nitty-gritty. Clin Microbiol Rev 20(4): 593–621

The manufacturer's authorised representative in the EU is Springer Nature Customer Service Centre GmbH, Europaplatz 3, 69115 Heidelberg, Germany. If you have any concerns regarding our products, please contact ProductSafety@springernature.com

Printed and bound by CPI Group (UK) Ltd, Croydon, CR0 4YY
30/09/2026
02225309-0001